Elements of the Theory of Functions and Functional Analysis

VOLUMES 1 AND 2
TWO VOLUMES BOUND AS ONE

BY

A.N. KOLMOGOROV

AND

S.V. FOMIN

DOVER PUBLICATIONS, INC.
Mineola, New York

Bibliographical Note

This Dover edition, first published in 1999, is an unabridged and unaltered republication in one volume of the work first published in two volumes by Graylock Press, Rochester, New York. Translated from the first (1954 and 1960) Russian editions by Leo F. Boron (Volume 1) and Hyman Kamel and Horace Komm (Volume 2).

International Standard Book Number: 0-486-40683-0

Manufactured in the United States by Courier Corporation
40683009
www.doverpublications.com

Elements of the Theory of Functions and Functional Analysis

Volume 1: Metric and Normed Spaces

CONTENTS

ADDENDUM TO CHAPTER III

Generalized Functions

CHAPTER IV

Linear Operator Equations

PREFACE

The present book is a revised version of material given by the authors in two courses at the Moscow State University. A course in functional analysis which included, on the one hand, basic information about the theory of sets, measure and the Lebesgue integral and, on the other hand, examples of the applications of the general methods of set theory, theory of functions of a real variable, and functional analysis to concrete problems in classical analysis (e.g. the proof of existence theorems, the discussion of integral equations as an example of a special case of the theory of operator equations in linear spaces, etc.) was given by the first author, A. N. Kolmogorov, in the Department of Mathematics and Mechanics. A somewhat less comprehensive course was given by the second author, S. V. Fomin, for students specializing in mathematics and theoretical physics in the Department of Physics. A. A. Petrov's notes of A. N. Kolmogorov's lectures were used in a number of sections.

The material included in the first volume is clear from the table of contents. The theory of measure and the Lebesgue integral, Hilbert space, theory of integral equations with symmetric kernel and orthogonal systems of functions, the elements of nonlinear functional analysis, and some applications of the methods of functional analysis to problems arising in the mathematics of numerical methods will be considered in later volumes.

A. KOLMOGOROV
S. FOMIN

February 1954

TRANSLATOR'S NOTE

This volume is a translation of A. N. Kolmogorov and S. V. Fomin's *Elementy Teorii Funkcii i Funkcional'nogo Analiza., I. Metriceskie i Normirovannye Prostranstva*. Chapter I is a brief introduction to set theory and mappings. There is a clear presentation of the elements of the theory of metric and complete metric spaces in Chapter II. The latter chapter also has a discussion of the principle of contraction mappings and its applications to the proof of existence theorems in the theory of differential and integral equations. The material on continuous curves in metric spaces is not usually found in textbook form. The elements of the theory of normed linear spaces are taken up in Chapter III where the Hahn-Banach theorem is proved for real separable normed linear spaces. The reader interested in the extension of this theorem to complex linear spaces is referred to the paper *Extension of Functionals on Complex Linear Spaces*, Bulletin AMS, 44 (1938), 91–93, by H. F. Bohnenblust and A. Sobczyk. Chapter III also deals with weak sequential convergence of elements and linear functionals and gives a discussion of adjoint operators. The addendum to Chapter III discusses Sobolev's work on generalized functions which was later generalized further by L. Schwartz. The main results here are that every generalized function has derivatives of all orders and that every convergent series of generalized functions can be differentiated term by term any number of times. Chapter IV, on linear operator equations, discusses spectra and resolvents for continuous linear operators in a complex Banach space. Minor changes have been made, notably in the arrangement of the proof of the Hahn-Banach theorem. A bibliography, listing basic books covering the material in this volume, was added by the translator. Lists of symbols, definitions and theorems have also been added at the end of the volume for the convenience of the reader.

Milwaukee 1957   *Leo F. Boron*

Chapter I

FUNDAMENTAL CONCEPTS OF SET THEORY

§1. The concept of set. Operations on sets

In mathematics as in everyday life we encounter the concept of set. We can speak of the set of faces of a polyhedron, students in an auditorium, points on a straight line, the set of natural numbers, and so on. The concept of a set is so general that it would be difficult to give it a definition which would not reduce to simply replacing the word "set" by one of the equivalent expressions: aggregate, collection, etc.

The concept of set plays an extraordinarily important role in modern mathematics not only because the theory of sets itself has become at the present time a very extensive and comprehensive discipline but mainly because of the influence which the theory of sets, arising at the end of the last century, exerted and still exerts on mathematics as a whole. Here we shall briefly discuss only those very basic set-theoretic concepts which will be used in the following chapters. The reader will find a significantly more detailed exposition of the theory of sets in, for example, the books by P. S. Aleksandrov: *Introduction to the General Theory of Sets and Functions*, where he will also find a bibliography for further reading, E. Kamke: *Theory of Sets*, F. Hausdorff: *Mengenlehre*, and A. Fraenkel: *Abstract Set Theory*.

We shall denote sets by upper case letters A, B, $\cdots$ and their elements by lower case letters a, b, $\cdots$. The statement "the element a belongs to the set A" will be written symbolically as $a \in A$; the expression $a \notin A$ means that the element a does not belong to the set A. If all the elements of which the set A consists are also contained in the set B (where the case $A = B$ is not excluded), then A will be called a subset of B and we shall write $A \subseteq B$. (The notation $A \subset B$ denotes that A is a subset of the set B and $A \neq B$, i. e. there exists at least one element in B which does not belong to A. A is then said to be a *proper* subset of the set B.) For example, the integers form a subset of the set of real numbers.

Sometimes, in speaking about an arbitrary set (for example, about the set of roots of a given equation) we do not know in advance whether or not this set contains even one element. For this reason it is convenient to introduce the concept of the so-called void set, that is, the set which does not contain any elements. We shall denote this set by the symbol θ. Every set contains θ as a subset.

If A and B are arbitrary sets, then their *sum* or *union* is the set $C = A \cup B$ consisting of all elements which belong to at least one of the sets A and B (Fig. 1).

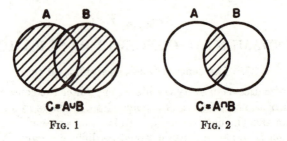

FIG. 1   FIG. 2

We define the sum of an arbitrary (finite or infinite) number of sets analogously: if A_α are arbitrary sets, then their sum $A = \bigcup_\alpha A_\alpha$ is the totality of elements each of which belongs to at least one of the sets A_α.

The *intersection* of two sets A and B is the set $C = A \cap B$ which consists of all the elements belonging to both A and B (Fig. 2). For example, the intersection of the set of all even integers and the set of all integers which are divisible by three is the same as the set of all integers which are divisible by six. The intersection of an arbitrary (finite or infinite) number of sets A_α is the set $A = \bigcap_\alpha A_\alpha$ of all elements which belong to all of the sets A_α. If $A \cap B = \theta$, we shall say that A and B are disjoint. The same term will apply to any collection of sets $\{A_\alpha\}$ for which $A_\beta \cap A_\gamma = \theta$, $\beta \neq \gamma$.

The operations of union and intersection are connected by the following relations:

$$(1) \qquad (A \cup B) \cap C = (A \cap C) \cup (B \cap C),$$

$$(2) \qquad (A \cap B) \cup C = (A \cup C) \cap (B \cup C).$$

We shall verify the first of these two relations. Let the element x belong to the set on the left side of equation (1). This means that x belongs to C and moreover that it belongs to at least one of the sets A and B. But then x belongs to at least one of the sets $A \cap C$ and $B \cap C$, i.e. it belongs to the right member of equation (1). To prove the converse, let $x \in (A \cap C) \cup (B \cap C)$. Then $x \in A \cap C$ and/or $x \in B \cap C$. Consequently, $x \in C$ and, moreover, x belongs to at least one of the sets A and B, i. e. $x \in (A \cup B)$. Thus we have shown that $x \in (A \cup B) \cap C$. Hence, equation (1) has been verified. Equation (2) is verified analogously.

We define further the operation of subtraction for sets. The *difference* of the sets A and B is the set $C = A \setminus B$ of those elements in A which are not contained in B (Fig. 3). In general it is not assumed here that $A \supset B$.

In some instances, for example in the theory of measure, it is convenient to consider the so-called *symmetric difference* of two sets A and B; the symmetric difference is defined as the sum of the differences $A \setminus B$ and $B \setminus A$ (Fig. 4). We shall denote the symmetric difference of the sets A and B by

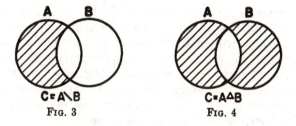

Fig. 3   Fig. 4

the symbol $A \Delta B$. Its definition is written symbolically as follows:

$$A \Delta B = (A \setminus B) \cup (B \setminus A).$$

EXERCISE. Show that $A \Delta B = (A \cup B) \setminus (A \cap B)$.

In the sequel we shall frequently have occasion to consider various sets all of which are subsets of some fundamental set S, such as for example various point sets on the real line. In this case the difference $S \setminus A$ is called the *complement* of the set A with respect to S.

In the theory of sets and in its applications a very important role is played by the so-called principle of duality which is based on the following two relations.

1. *The complement of a sum is equal to the intersection of the complements,*

(3) $$S \setminus \bigcup_\alpha A_\alpha = \bigcap_\alpha (S \setminus A_\alpha).$$

2. *The complement of an intersection is equal to the sum of the complements,*

(4) $$S \setminus \bigcap_\alpha A_\alpha = \bigcup_\alpha (S \setminus A_\alpha).$$

By virtue of these relations, we can start with an arbitrary theorem concerning a system of subsets of a fixed set S and automatically obtain the dual theorem by replacing the sets under consideration by their complements, sums by intersections, and intersections by sums. Theorem 1′, Chapter II, §10, is an example of the application of this principle.

We shall now prove relation (3). Let $x \in S \setminus \bigcup_\alpha A_\alpha$. This means that x does not belong to the sum $\bigcup_\alpha A_\alpha$, i.e. x does not belong to any of the sets A_α. Consequently, x belongs to each of the complements $S \setminus A_\alpha$ and therefore $x \in \bigcap_\alpha (S \setminus A_\alpha)$. Conversely, let $x \in \bigcap_\alpha (S \setminus A_\alpha)$, i.e. x belongs to every $S \setminus A_\alpha$. Consequently, x does not belong to any of the sets A_α, i.e. it does not belong to their sum $\bigcup_\alpha A_\alpha$. But then $x \in S \setminus \bigcup_\alpha A_\alpha$; this concludes the proof of (3). We prove relation (4) analogously.

§2. Finite and infinite sets. Denumerability

In considering various sets we note that for some of them we can indicate the number of elements in the set, if not actually then at least in theory. Such, for example, is the set of chairs in a given room, the set of pencils in a box, the set of all automobiles in a given city, the set of all molecules of water on the earth, and so on. Each of these sets contains a finite number of elements, although the number may not be known to us. On the other hand, there exist sets consisting of an infinite number of elements. Such, for example, are the sets of all natural numbers, all points on the real line, all circles in the plane, all polynomials with rational coefficients, and so forth. In this connection, when we say that a set is infinite, we have in mind that we can remove one element, two elements, and so on, where after each step elements still remain in the set.

When we consider two finite sets it may occur that the number of elements in both of them is the same or it may occur that in one of these sets the number of elements is greater than in the other, i.e. we can compare finite sets by means of the number of elements they contain. The question can be asked whether or not it is possible in a similar way to compare infinite sets. In other words, does it make sense, for example, to ask which of the following sets is the larger: the set of circles in the plane or the set of rational points on the real line, the set of functions defined on the segment [0, 1] or the set of straight lines in space, and so on?

Consider more carefully how we compare two finite sets. We can tackle the problem in two ways. We can either count the number of elements in each of the two sets and thus compare the two sets or we can try to establish a correspondence between the elements of these sets by assigning to each element of one of the sets one and only one element of the other set, and conversely; such a correspondence is said to be one-to-one. Clearly, a one-to-one correspondence between two finite sets can be established if and only if the number of elements in both sets is the same. For instance, in order to verify that the number of students in a group and the number of seats in an auditorium are the same, rather than counting each of the sets, one can seat each of the students in a definite seat. If there is a sufficient number of seats and no seat remains vacant, i.e. if a one-to-one correspondence is set up between these two sets, then this will mean that the number of elements is the same in both.

But it is obvious that the first method (counting the number of elements) is suitable only for comparing finite sets, whereas the second method (setting up a one-to-one correspondence) is suitable to the same degree for infinite as well as for finite sets.

Among all possible infinite sets the simplest is the set of natural numbers. We shall call every set whose elements can be put into one-to-one correspondence with all the natural numbers a *denumerable set*. In other words, a denumerable set is a set whose elements can be indexed in the form of an infinite sequence: $a_1, a_2, \cdots, a_n, \cdots$. The following are examples of denumerable sets.

1. The set of all integers. We can establish a one-to-one correspondence

between the set of all integers and the set of all natural numbers in the following way:

$$\begin{matrix} 0 & -1 & 1 & -2 & 2 & \cdots \\ 1 & 2 & 3 & 4 & 5 & \cdots \end{matrix},$$

where in general we set $n \leftrightarrow 2n + 1$ if $n \geq 0$ and $n \leftrightarrow -2n$ if $n < 0$.

2. The set of all positive even integers. The correspondence is obviously $n \leftrightarrow 2n$.

3. The set of numbers $2, 4, 8, \cdots, 2^n, \cdots$. Assign to each number 2^n the corresponding n; this correspondence is obviously one-to-one.

4. We now consider a slightly more complicated example: we shall show that the set of all rational numbers is denumerable. Every rational number can be written in the form of an irreducible fraction $\alpha = p/q$, $q > 0$. Call the sum $n = |p| + q$ the *height* of the rational number α. It is clear that the number of fractions having height n is finite. For instance, the number $0/1 = 0$ is the only number having height 1; the numbers $1/1$ and $-1/1$ are the only numbers having height 2; the numbers $2/1$, $1/2$, $-2/1$ and $-1/2$ have height 3; and so forth. We enumerate all the rational numbers in the order of increasing heights, i.e. first the numbers with height 1, then the numbers with height 2, etc. This process assigns some index to each rational number, i.e. we shall have set up a one-to-one correspondence between the set of all natural numbers and the set of all rational numbers.

An infinite set which is not denumerable is said to be a *nondenumerable set*.

We establish some general properties of denumerable sets.

1°. *Every subset of a denumerable set is either finite or denumerable.*

Proof. Let A be a denumerable set and let B be a subset of A. If we enumerate the elements of the set A: $a_1, a_2, \cdots, a_n, \cdots$ and let n_1, $n_2, \cdots$ be the natural numbers which correspond to the elements in B in this enumeration, then if there is a largest one among these natural numbers, B is finite; in the other case B is denumerable.

2°. *The sum of an arbitrary finite or denumerable set of denumerable sets is again a finite or denumerable set.*

Proof. Let $A_1, A_2, \cdots$ be denumerable sets. All their elements can be written in the form of the following infinite table:

$$\begin{matrix} a_{11} & a_{12} & a_{13} & a_{14} & \cdots \\ a_{21} & a_{22} & a_{23} & a_{24} & \cdots \\ a_{31} & a_{32} & a_{33} & a_{34} & \cdots \\ a_{41} & a_{42} & a_{43} & a_{44} & \cdots \\ \cdots & \cdots & \cdots & \cdots & \cdots \end{matrix},$$

where the elements of the set A_1 are listed in the first row, the elements of A_2 are listed in the second row, and so on. We now enumerate all these elements by the "diagonal method", i.e. we take a_{11} for the first element, a_{12} for the second, a_{21} for the third, and so forth, taking the elements in the order indicated by the arrows in the following table:

$$
\begin{array}{cccc}
a_{11} \rightarrow a_{12} & a_{13} \rightarrow a_{14} \cdots \\
\swarrow \quad \nearrow \quad \swarrow \\
a_{21} \quad a_{22} \quad a_{23} \quad a_{24} \cdots \\
\downarrow \nearrow \quad \swarrow \\
a_{31} \quad a_{32} \quad a_{33} \quad a_{34} \cdots \\
\swarrow \\
a_{41} \quad a_{42} \quad a_{43} \quad a_{44} \cdots \\
\cdots \cdots \cdots \cdots \cdots \cdots \cdots \cdots \cdots \; .
\end{array}
$$

It is clear that in this enumeration every element of each of the sets A_i receives a definite index, i.e. we shall have established a one-to-one correspondence between all the elements of all the A_1, A_2, $\cdots$ and the set of natural numbers. This completes the proof of our assertion.

EXERCISES. 1. Prove that the set of all polynomials with rational coefficients is denumerable.

2. The number ξ is said to be algebraic if it is a zero of some polynomial with rational coefficients. Prove that the set of all algebraic numbers is denumerable.

3. Prove that the set of all rational intervals (i.e. intervals with rational endpoints) on the real line is denumerable.

4. Prove that the set of all points in the plane having rational coordinates is denumerable. *Hint*: Use Theorem 2°.

3°. *Every infinite set contains a denumerable subset.*

Proof. Let M be an infinite set and consider an arbitrary element a_1 in M. Since M is infinite, we can find an element a_2 in M which is distinct from a_1, then an element a_3 distinct from both a_1 and a_2, and so forth. Continuing this process (which cannot terminate in a finite number of steps since M is infinite), we obtain a denumerable subset $A = \{a_1, a_2, a_3, \cdots\}$ of the set M. This completes the proof of the theorem.

This theorem shows that denumerable sets are, so to speak, the "smallest" of the infinite sets. The question whether or not there exist nondenumerable infinite sets will be considered in §4.

§3. Equivalence of sets

We arrived at the concept of denumerable set by establishing a one-to-one correspondence between certain of infinite sets and the set of natural numbers; at the same time we gave a number of examples of denumerable sets and some of their general properties.

It is clear that by setting up a one-to-one correspondence it is possible not only to compare infinite sets with the set of natural numbers; it is possible to compare any two sets by this method. We introduce the following definition.

DEFINITION. Two sets M and N are said to be *equivalent* (notation: $M \sim N$) if a one-to-one correspondence can be set up between their elements.

The concept of equivalence is applicable to arbitrary sets, infinite as well as finite. It is clear that two finite sets are equivalent if, and only if, they consist of the same number of elements. The definition we introduced above of a denumerable set can now be formulated in the following way: *a set is said to be denumerable if it is equivalent to the set of natural numbers.*

EXAMPLES. 1. The sets of points on two arbitrary segments $[a, b]$ and $[c, d]$ are equivalent. A method for establishing a one-to-one correspondence between them is shown in Fig. 5. Namely, the points p and q correspond if they lie on the same ray emanating from the point O in which the straight lines ac and bd intersect.

2. The set of all points in the closed complex plane is equivalent to the set of all points on the sphere. A one-to-one correspondence $a \leftrightarrow z$ can be established, for example, with the aid of stereographic projection (Fig. 6).

3. The set of all real numbers in the interval $(0, 1)$ is equivalent to the set of all points on the real line. The correspondence can be established, for example, with the aid of the function

$$y = (1/\pi) \arctan x + \tfrac{1}{2}.$$

It is clear directly from the definition that two sets each equivalent to a third set are equivalent.

Considering the examples introduced here and in §2 one can make the following interesting deduction: in a number of cases an infinite set proves to be equivalent to a proper subset of itself. For example, there are "as many" natural numbers as there are integers or even as there are of all rationals; there are "as many" points on the interval $(0, 1)$ as there are on the entire real line, and so on. It is not difficult to convince oneself of the fact that this situation is characteristic of all infinite sets.

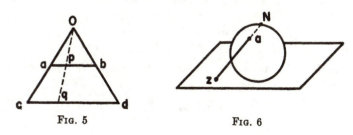

FIG. 5   FIG. 6

In fact, in §2 (Theorem 3°) we showed that every infinite set M has a denumerable subset; let this set be

$$A = \{a_1, a_2, \cdots, a_n, \cdots\}.$$

We partition A into two denumerable subsets

$$A_1 = \{a_1, a_3, a_5, \cdots\} \quad \text{and} \quad A_2 = \{a_2, a_4, a_6, \cdots\}.$$

Since A and A_1 are denumerable, a one-to-one correspondence can be set up between them. This correspondence can be extended to a one-to-one correspondence between the sets

$$A_1 \cup (M \setminus A) = M \setminus A_2 \quad \text{and} \quad A \cup (M \setminus A) = M$$

which assigns to each element in $M \setminus A$ this element itself. The set $M \setminus A_2$ is a proper subset of the set M. We thus obtain the following result:

Every infinite set is equivalent to some proper subset of itself.

This property can be taken as the definition of an infinite set.

EXERCISE. Prove that if M is an arbitrary infinite set and A is denumerable, then $M \sim M \cup A$.

§4. Nondenumerability of the set of real numbers

In §2 we introduced a number of examples of denumerable sets. The number of these examples could be significantly increased. Moreover, we showed that if we take sums of denumerable sets in finite or infinite number we again obtain denumerable sets. The following question arises naturally: do there exist in general nondenumerable infinite sets? The following theorem gives an affirmative answer to this question.

THEOREM. *The set of real numbers in the closed interval* [0, 1] *is nondenumerable.*

Proof. We shall assume the contrary, i.e., that all the real numbers lying on the segment [0, 1], each of which can be written in the form of an infinite decimal, can be arranged in the form of a sequence

(1)
$$
\begin{array}{llllll}
0.a_{11} & a_{12} & a_{13} & \cdots & a_{1n} & \cdots \\
0.a_{21} & a_{22} & a_{23} & \cdots & a_{2n} & \cdots \\
0.a_{31} & a_{32} & a_{33} & \cdots & a_{3n} & \cdots \\
\cdots & \cdots & \cdots & \cdots & \cdots & \cdots \\
0.a_{n1} & a_{n2} & a_{n3} & \cdots & a_{nn} & \cdots \\
\cdots & \cdots & \cdots & \cdots & \cdots & \cdots,
\end{array}
$$

where each a_{ik} is one of the numbers $0, 1, \cdots, 9$. We now construct the decimal

(2)
$$0.b_1 b_2 b_3 \cdots b_n \cdots$$

in the following manner: for b_1 we take an arbitrary digit which does not coincide with a_{11}, for b_2 an arbitrary digit which does not coincide with a_{22}, and so on; in general, for b_n we take an arbitrary digit not coinciding with a_{nn}. This decimal cannot coincide with any of the decimals appearing in Table (1). In fact, it differs from the first decimal of Table (1) at least in the first digit by construction, from the second decimal in the second digit and so forth; in general, since $b_n \neq a_{nn}$ for all n, decimal (2) cannot coincide with any of the decimals appearing in Table (1). Thus, the assumption that there is some way of enumerating all the real numbers lying on the segment [0, 1] has led to a contradiction.

The above proof lacks precision. Namely, some real numbers can be written in the form of a decimal in two ways: in one of them there is an infinite number of zeros and in the other an infinite number of nines; for example,

$$\tfrac{1}{2} = 0.5000 \cdots = 0.4999 \cdots .$$

Thus, the noncoincidence of two decimals still does not mean that these decimals represent distinct numbers.

However, if decimal (2) is constructed so that it contains neither zeros nor nines, then setting, for example, $b_n = 2$, if $a_{nn} = 1$ and $b_n = 1$ if $a_{nn} \neq 1$, the above-mentioned objection is avoided.

So we have found an example of a nondenumerable infinite set. We shall point out some examples of sets which are equivalent to the set of real numbers in the closed interval [0, 1].

1. The set of all points on an arbitrary segment $[a, b]$ or the points of the open interval (a, b).

2. The set of all points on a straight line.

3. The set of all points in the plane, in space, on the surface of a sphere, the points lying in the interior of a sphere, and so forth.

4. The set of all straight lines in the plane.

5. The set of all continuous functions of one or several variables.

In Examples 1 and 2 the proof offers no difficulty (see Examples 1 and 3, §3). In the other examples a direct proof is somewhat complicated.

EXERCISE. Using the results of this section and Exercise 2, §2, prove the existence of transcendental numbers, i.e., of real numbers which are not algebraic.

§5. The concept of cardinal number

If two finite sets are equivalent, they consist of the same number of elements. If M and N are two arbitrary equivalent sets we say that M and N have the *same cardinal number* (or the *same power* or the *same potency*). Thus, cardinal number is what all equivalent sets have in common. For finite sets the concept of cardinal number coincides simply with the concept of number of elements in the set. The cardinal number of the set of natural numbers (i.e. of any denumerable set) is denoted by the symbol $\aleph_0$ (read "aleph zero"). Sets which are equivalent to the set of real numbers between 0 and 1 are said to be sets having the power of the continuum. This power is denoted by the symbol c.

As a rule, all infinite sets which are encountered in analysis are either denumerable or have cardinal number c.

If the set A is equivalent to some subset of the set B but is not equivalent to the entire set B, then we say that the cardinal number of the set A is less than the cardinal number of the set B.

Logically, besides the two possibilities indicated, namely: 1) A equivalent to B and 2) A equivalent to a subset of B but not equivalent to all of B, we shall allow two more: 3) A is equivalent to some subset of B and B is equivalent to some subset of A; 4) A and B are not equivalent and in neither of these sets is there a subset equivalent to the other set. It is possible to show that in Case 3 the sets A and B are themselves equivalent (this is the Cantor-Bernstein theorem) but that Case 4 is in fact impossible (Zermelo's theorem); however, we shall not give the proof of these two rather complicated theorems here (see, for example, P. S. Aleksandrov: *Introduction to the General Theory of Sets and Functions*, Chapter I, §6 and Chapter II, §6; E. Kamke: *Theory of Sets*; F. Hausdorff: *Mengenlehre*; and A. Fraenkel: *Abstract Set Theory*).

As was pointed out at the end of §2 denumerable sets are the "smallest" infinite sets. In §4 we showed that there exist infinite sets whose infiniteness is of a higher order; these were sets having the cardinal number of the continuum. But do there exist infinite cardinal numbers exceeding the cardinal number of the continuum? In general, does there exist some "highest" cardinal number or not? It turns out that the following theorem is true.

THEOREM. *Let M be a set of cardinal number* $\mathfrak{m}$. *Further, let $\mathfrak{M}$ be the set whose elements are all possible subsets of the set M. Then $\mathfrak{M}$ has greater cardinal number than the cardinal number* $\mathfrak{m}$ *of the initial set M.*

Proof. It is easy to see that the cardinal number of the set $\mathfrak{M}$ cannot be less than the cardinal number $\mathfrak{m}$ of the initial set; in fact, those subsets of M each of which consists of only one element form a subset of $\mathfrak{M}$ which is equivalent to the set M. It remains to prove that these cardinal numbers cannot coincide. Let us assume the contrary; then $\mathfrak{M}$ and M are equivalent and we can set up a one-to-one correspondence between them. Let $a \leftrightarrow A$, $b \leftrightarrow B$, $\cdots$ be a one-to-one correspondence between the elements of the set M and all of its subsets, i.e., the elements of the set $\mathfrak{M}$. Now let X be the set of elements in M which do not belong to those subsets to which they correspond (for example, if $a \in A$ then $a \notin X$, if $b \notin B$ then $b \in X$, and so forth). X is a subset of M, i.e. it is some element in $\mathfrak{M}$. By assumption X must correspond to some element $x \in M$. Let us see whether or not the element x belongs to the subset X. Let us assume $x \notin X$. But by definition X consists of all those elements which are not contained in the subset to which they correspond and consequently the element x ought to be included in X. Conversely, if we assume that x belongs to X, we may conclude that x cannot belong to X since X contains only those elements which do not belong to the subset to which they correspond. Thus the element corresponding to the subset X ought simultaneously to be contained in and not contained in X. This implies that in general such an element does not exist, i.e. that it is impossible to establish a one-to-one correspondence between the elements of the set M and all its subsets. This completes the proof of the theorem.

Thus for an arbitrary cardinal number we can in reality construct a set of greater cardinal number and then a still greater cardinal number, and so on, obtaining in this way a hierarchy of cardinal numbers which is not bounded in any way.

Exercise. Prove that the set of all numerical functions defined on a set M has a greater cardinal number than the cardinal number of the set M. *Hint*: Use the fact that the set of all characteristic functions (i.e. functions assuming only the values 0 and 1) defined on M is equivalent to the set of all subsets of M.

§6. Partition into classes

The reader can omit this section on a first reading and return to it in the sequel for information as required.

In the most varied questions it occurs that we encounter partitions of a set into disjoint subsets. For example, the plane (considered as a set of points) can be partitioned into straight lines parallel to the y-axis, three-dimensional space can be represented as the set of all concentric spheres of different radii, the inhabitants of a given city can be partitioned into groups according to their year of birth, and so forth.

If a set M is represented in some manner as a sum of disjoint subsets, we speak of a partitioning of the set M into classes.

Ordinarily we encounter partitions which are obtained by means of indicating some rule according to which the elements of the set M are combined into classes. For example, the set of all triangles in the plane can be partitioned into classes of triangles which are congruent to one another or triangles which have the same area; all polynomials in x can be partitioned into classes by collecting all polynomials having the same zeros into one class, and so on.

Rules according to which the elements of a set are partitioned into classes can be of the most varied sort. But of course all these rules cannot be entirely arbitrary. Let us assume, for example, that we should like to partition all real numbers into classes by including the number b in the same class as the number a if, and only if, $b > a$. It is clear that no partition of the real numbers into classes can be obtained in this way because if $b > a$, i.e. if b must be included in the same class as a, then $a < b$, i.e. the number a *must not* belong to the same class as b. Moreover, since a is not larger than a, then a ought not belong in the class which contains it! We consider another example. We shall see whether or not it is possible to partition all inhabitants of a given city into classes by putting two persons into the same class if, and only if, they are acquaintances. It is clear that such a partition cannot be realized because if A is an acquaintance of B and B is an acquaintance of C, then this does not at all mean that A is an acquaintance of C. Thus, if we put A into the same class as B and B into the same class as C, it may follow that two persons A and C who are not acquaintances are in the same class. We obtain an analogous result if we attempt to partition the points of the plane into classes so that those and only those points whose mutual distance does not exceed 1 are put into one class.

The examples introduced above point out those conditions which must be satisfied by any rule if it is to realize a partition of the elements of a set into classes.

Let M be a set and let some pair (a, b) of elements of this set be "marked." [Here the elements a and b are taken in a definite order, i.e. (a, b) and (b, a) are two distinct pairs.] If (a, b) is a "marked" pair, we shall say that the element a is related to b by the relation φ and we shall denote this fact by means of the symbol $a \varphi b$. For example, if we wish to partition triangles into classes of triangles having the same area, then $a \varphi b$ is to mean: "triangle a has the same area as triangle b." We shall say that the given relation is an equivalence relation if it possesses the following properties:

1. *Reflexivity*: $a \varphi a$ for any element $a \in M$;
2. *Symmetry*: if $a \varphi b$, then necessarily $b \varphi a$;
3. *Transitivity*: if $a \varphi b$ and $b \varphi c$, then $a \varphi c$.

Obviously every partition of a given set into classes defines some equivalence relation among the elements of this set.

In fact, if $a \varphi b$ means "a belongs to the same class as b", then this relation will be reflexive, symmetric and transitive, as is easy to verify.

Conversely, if $a \varphi b$ is an equivalence relation between the elements of the set M, then putting into one class those and only those elements which are equivalent we obtain a partition of the set M into classes.

In fact, let K_a be the class of elements in M which are equivalent to a fixed element a. By virtue of the property of reflexivity the element a itself belongs to the class K_a. We shall show that two classes K_a and K_b either coincide or are disjoint. Let an element c belong simultaneously to K_a and to K_b, i.e. $c \varphi a$ and $c \varphi b$.

Then by virtue of symmetry, $a \varphi c$, and by virtue of transitivity,

(1) $$a \varphi b.$$

Now if x is an arbitrary element in K_a, i.e. $x \varphi a$, then by virtue of (1) and the transitivity property, $x \varphi b$, i.e. $x \in K_b$.

Conversely, if y is an arbitrary element in K_b, i.e. if $y \varphi b$, then by virtue of relation (1) which can be written in the form $b \varphi a$ (symmetry!) and the transitivity property, $y \varphi a$, i.e. $y \in K_a$. Thus, two classes K_a and K_b having at least one element in common coincide.

We have in fact obtained a partition of the set M into classes according to the given equivalence relation.

§7. Mappings of sets. General concept of function

In analysis the concept of function is introduced in the following way. Let X be a set on the real line. We say that a function f is defined on this set if to each number $x \in X$ there is made to correspond a definite number $y = f(x)$. In this connection X is said to be the domain of the given function and Y, the set of all values assumed by this function, is called its range.

Now if instead of sets of numbers we consider sets of a completely arbitrary nature, we arrive at the most general concept of function, namely: Let M and N be two arbitrary sets; then we say that a function f is defined on M and assumes its values in N if to each element $x \in M$ there is made to correspond one and only one element in N. In the case of sets of an arbitrary nature instead of the term "function" we frequently use the term "mapping" and speak of a mapping of one set into another.

If a is any element in M, the element $b = f(a)$ in N which corresponds to it is called the *image* of the element a (under the mapping f). The set of all those elements in M whose image is a given element $b \in N$ is called the *inverse image* (or more precisely the *complete inverse image*) of the element b and is denoted by $f^{-1}(b)$.

If A is any set in M, the set of all elements of the form $\{f(a) : a \in A\}$ is called the image of A and is denoted by $f(A)$. In its turn, for every set B in N there is defined its inverse image $f^{-1}(B)$, namely $f^{-1}(B)$ is the set of all those elements in M whose images belong to B.

In this section we shall limit ourselves to the consideration of the most general properties of mappings.

We shall use the following terminology. We say that f is a mapping of the set M *onto* the set N if $f(M) = N$; in the general case, i.e. when $f(M) \subsetneq N$ we say that f is a mapping of M *into* N.

We shall establish the following properties of mappings.

THEOREM 1. *The inverse image of the sum of two sets is equal to the sum of their inverse images:*

$$f^{-1}(A \cup B) = f^{-1}(A) \cup f^{-1}(B).$$

Proof. Let the element x belong to the set $f^{-1}(A \cup B)$. This means that $f(x) \in A \cup B$, i.e. $f(x) \in A$ or $f(x) \in B$. But then x belongs to at least one of the sets $f^{-1}(A)$ and $f^{-1}(B)$, i.e. $x \in f^{-1}(A) \cup f^{-1}(B)$. Conversely, if $x \in f^{-1}(A) \cup f^{-1}(B)$, then x belongs to at least one of the sets $f^{-1}(A)$ and $f^{-1}(B)$, i.e. $f(x)$ belongs to at least one of the sets A and B and consequently $f(x) \in A \cup B$, whence it follows that $x \in f^{-1}(A \cup B)$.

THEOREM 2. *The inverse image of the intersection of two sets is equal to the intersection of their inverse images:*

$$f^{-1}(A \cap B) = f^{-1}(A) \cap f^{-1}(B).$$

Proof. If $x \in f^{-1}(A \cap B)$, then $f(x) \in A \cap B$, i.e. $f(x) \in A$ and $f(x) \in B$; consequently, $x \in f^{-1}(A)$ and $x \in f^{-1}(B)$, i.e. $x \in f^{-1}(A) \cap f^{-1}(B)$.

Conversely, if $x \in f^{-1}(A) \cap f^{-1}(B)$, i.e. $x \in f^{-1}(A)$ and $x \in f^{-1}(B)$, then $f(x) \in A$ and $f(x) \in B$, or in other words $f(x) \in A \cap B$.

Consequently $x \in f^{-1}(A \cap B)$.

Theorems 1 and 2 remain valid also for the sum and intersection of an arbitrary finite or infinite number of sets.

Thus, if in N some system of sets closed with respect to the operations of addition and taking intersections is selected, then their inverse images in M form a system which is likewise closed with respect to these operations.

THEOREM 3. *The image of the sum of two sets is equal to the sum of their images:*

$$f(A \cup B) = f(A) \cup f(B).$$

Proof. If $y \in f(A \cup B)$, then $y = f(x)$, where x belongs to at least one of the sets A and B. Consequently, $y = f(x) \in f(A) \cup f(B)$. Conversely, if $y \in f(A) \cup f(B)$ then $y = f(x)$, where x belongs to at least one of the sets A and B, i.e. $x \in A \cup B$ and consequently $y = f(x) \in f(A \cup B)$.

We note that in general the image of the intersection of two sets does not coincide with the intersection of their images. For example, let the mapping considered be the projection of the plane onto the x-axis. Then the segments

$$0 \le x \le 1; \quad y = 0,$$
$$0 \le x \le 1; \quad y = 1$$

do not intersect, but at the same time their images coincide.

The concept of mapping of sets is closely related to the concept of partitioning considered in the preceding section.

Let f be a mapping of the set A into the set B. If we collect into one class all those elements in A whose images in B coincide, we obviously obtain a partition of the set A. Conversely, let us consider an arbitrary set A and an arbitrary partitioning of A into classes. Let B be the totality of those classes into which the set A is partitioned. If we correspond to each element $a \in A$ that class (i.e. that element in B) to which a belongs, we obtain a mapping of the set A into the set B.

EXAMPLES. 1. Consider the projection of the xy-plane onto the x-axis. The inverse images of the points of the x-axis are vertical lines. Consequently to this mapping there corresponds a partitioning of the plane into parallel straight lines.

2. Subdivide all the points of three-dimensional space into classes by combining into one class the points which are equidistant from the origin of coordinates. Thus, every class can be represented by a sphere of some radius. A realization of the totality of all these classes is the set of all points lying on the ray $(0, \infty)$. So, to the partitioning of three-dimensional space into concentric spheres there corresponds the mapping of this space onto a half-line.

3. Combine all real numbers having the same fractional part into one class. The mapping corresponding to this partition is represented by the mapping of the real line onto a circle.

Chapter II

METRIC SPACES

§8. Definition and examples of metric spaces

Passage to the limit is one of the most important operations in analysis. The basis of this operation is the fact that the distance between any two points on the real line is defined. A number of fundamental facts from analysis are not connected with the algebraic nature of the set of real numbers (i.e. with the fact that the operations of addition and multiplication, which are subject to known laws, are defined for real numbers), but depend only on those properties of real numbers which are related to the concept of distance. This situation leads naturally to the concept of "metric space" which plays a fundamental role in modern mathematics. Further on we shall discuss the basic facts of the theory of metric spaces. The results of this chapter will play an essential role in all the following discussion.

DEFINITION. A *metric space* is the pair of two things: a set X, whose elements are called points, and a distance, i.e. a single-valued, nonnegative, real function $\rho(x, y)$, defined for arbitrary x and y in X and satisfying the following conditions:

1) $\rho(x, y) = 0$ if and only if $x = y$,
2) (axiom of symmetry) $\rho(x, y) = \rho(y, x)$,
3) (triangle axiom) $\rho(x, y) + \rho(y, z) \geq \rho(x, z)$.

The metric space itself, i.e. the pair X and ρ, will usually be denoted by $R = (X, \rho)$.

In cases where no misunderstanding can arise we shall sometimes denote the metric space by the same symbol X which is used for the set of points itself.

We list a number of examples of metric spaces. Some of the spaces listed below play a very important role in analysis.

1. If we set

$$\rho(x, y) = \begin{cases} 0, \text{ if } x = y, \\ 1, \text{ if } x \neq y, \end{cases}$$

for elements of an arbitrary set, we obviously obtain a metric space.

2. The set D^1 of real numbers with the distance function

$$\rho(x, y) = |x - y|$$

forms the metric space R^1.

3. The set D^n of ordered n-tuples of real numbers $x = (x_1, x_2, \cdots, x_n)$ with distance function

$$\rho(x, y) = \{\textstyle\sum_{k=1}^{n} (y_k - x_k)^2\}^{\frac{1}{2}}$$

is called Euclidean n-space R^n. The validity of Axioms 1 and 2 for R^n is obvious. To prove that the triangle axiom is also verified in R^n we make use of the Schwarz inequality

$$(1) \qquad \left(\sum_{k=1}^{n} a_k b_k\right)^2 \leq \sum_{k=1}^{n} a_k^2 \sum_{k=1}^{n} b_k^2.$$

(The Schwarz inequality follows from the identity

$$\left(\sum_{k=1}^{n} a_k b_k\right)^2 = \left(\sum_{k=1}^{n} a_k^2\right)\left(\sum_{k=1}^{n} b_k^2\right) - \tfrac{1}{2}\sum_{i=1}^{n}\sum_{j=1}^{n}(a_i b_j - b_i a_j)^2,$$

which can be verified directly.) If

$$x = (x_1, x_2, \cdots, x_n), \quad y = (y_1, y_2, \cdots, y_n) \quad \text{and} \quad z = (z_1, z_2, \cdots, z_n),$$

then setting

$$y_k - x_k = a_k, \qquad z_k - y_k = b_k,$$

we obtain

$$z_k - x_k = a_k + b_k;$$

by the Schwarz inequality

$$\sum_{k=1}^{n}(a_k + b_k)^2 = \sum_{k=1}^{n} a_k^2 + 2\sum_{k=1}^{n} a_k b_k + \sum_{k=1}^{n} b_k^2$$
$$\leq \sum_{k=1}^{n} a_k^2 + 2\left\{\sum_{k=1}^{n} a_k^2 \sum_{k=1}^{n} b_k^2\right\}^{\frac{1}{2}} + \sum_{k=1}^{n} b_k^2$$
$$= \left[\left(\sum_{k=1}^{n} a_k^2\right)^{\frac{1}{2}} + \left(\sum_{k=1}^{n} b_k^2\right)^{\frac{1}{2}}\right]^2,$$

i.e.

$$\rho^2(x, z) \leq \{\rho(x, y) + \rho(y, z)\}^2$$

or

$$\rho(x, z) \leq \rho(x, y) + \rho(y, z).$$

4. Consider the space R_0^n in which the points are again ordered n-tuples of numbers $(x_1, x_2, \cdots, x_n)$, and for which the distance function is defined by the formula

$$\rho_0(x, y) = \max\{|y_k - x_k|; 1 \leq k \leq n\}.$$

The validity of Axioms 1–3 is obvious. In many questions of analysis this space is no less suitable than Euclidean space R^n.

Examples 3 and 4 show that sometimes it is actually important to have different notations for the set of points of a metric space and for the metric space itself because the same point set can be metrized in various ways.

5. The set $C[a, b]$ of all continuous real-valued functions defined on the segment $[a, b]$ with distance function

$$(2) \qquad \rho(f, g) = \sup\{|g(t) - f(t)|; a \leq t \leq b\}$$

likewise forms a metric space. Axioms 1–3 can be verified directly. This space plays a very important role in analysis. We shall denote it by the same symbol $C[a, b]$ as the set of points of this space. The space of continuous functions defined on the segment $[0, 1]$ with the metric given above will be denoted simply by C.

6. We denote by l_2 the metric space in which the points are all possible sequences $x = (x_1, x_2, \cdots, x_n, \cdots)$ of real numbers which satisfy the condition $\sum_{k=1}^{\infty} x_k^2 < \infty$ and for which the distance is defined by means of the formula

$$(3) \qquad \rho(x, y) = \{ \sum_{k=1}^{\infty} (y_k - x_k)^2 \}^{\frac{1}{2}}.$$

We shall first prove that the function $\rho(x, y)$ defined in this way always has meaning, i.e. that the series $\sum_{k=1}^{\infty} (y_k - x_k)^2$ converges. We have

$$(4_n) \qquad \{ \sum_{k=1}^{n} (y_k - x_k)^2 \}^{\frac{1}{2}} \leq (\sum_{k=1}^{n} x_k^2)^{\frac{1}{2}} + (\sum_{k=1}^{n} y_k^2)^{\frac{1}{2}}$$

for arbitrary natural number n (see Example 3).

Now let $n \to \infty$. By hypothesis, the right member of this inequality has a limit. Thus, the expression on the left is bounded and does not decrease as $n \to \infty$; consequently, it tends to a limit, i.e. formula (3) has meaning. Replacing x by $-x$ in (4_n) and passing to the limit as $n \to \infty$, we obtain

$$(4) \qquad \{ \sum_{k=1}^{\infty} (y_k + x_k)^2 \}^{\frac{1}{2}} \leq (\sum_{k=1}^{\infty} x_k^2)^{\frac{1}{2}} + (\sum_{k=1}^{\infty} y_k^2)^{\frac{1}{2}};$$

but this is essentially the triangle axiom. In fact, let

$$a = (a_1, a_2, \cdots, a_n, \cdots),$$
$$b = (b_1, b_2, \cdots, b_n, \cdots),$$
$$c = (c_1, c_2, \cdots, c_n, \cdots)$$

be three points in l_2. If we set

$$b_k - a_k = x_k, \qquad c_k - b_k = y_k,$$

then

$$c_k - a_k = y_k + x_k$$

and, by virtue of (4),

$$\{ \sum_{k=1}^{\infty} (c_k - a_k)^2 \}^{\frac{1}{2}} \leq \{ \sum_{k=1}^{\infty} (b_k - a_k)^2 \}^{\frac{1}{2}} + \{ \sum_{k=1}^{\infty} (c_k - b_k)^2 \}^{\frac{1}{2}},$$

i.e.

$$\rho(a, c) \leq \rho(a, b) + \rho(b, c).$$

7. Consider, as in Example 5, the totality of all continuous functions on the segment $[a, b]$, but now let the distance be defined by setting

$$(5) \qquad \rho(x, y) = \left[\int_a^b \{ x(t) - y(t) \}^2 \, dt \right]^{\frac{1}{2}}.$$

This metric space is denoted by $C^2[a, b]$ and is called the *space of continuous functions with quadratic metric*. Here again Axioms 1 and 2 in the definition of a metric space are obvious and the triangle axiom follows immediately from the Schwarz inequality

$$\left\{\int_a^b x(t)y(t)\, dt\right\}^2 \le \int_a^b x^2(t)\, dt \int_a^b y^2(t)\, dt,$$

which can be obtained, for instance, from the following easily-verified identity:

$$\left\{\int_a^b x(t)y(t)\, dt\right\}^2 = \int_a^b x^2(t)\, dt \int_a^b y^2(t)\, dt$$

$$- \frac{1}{2}\int_a^b\int_a^b [x(s)y(t) - y(s)x(t)]^2\, ds\, dt.$$

8. Consider the set of all bounded sequences $x = (x_1, x_2, \cdots, x_n, \cdots)$ of real numbers. We obtain the metric space M^∞ if we set

(6) $$\rho(x, y) = \sup |y_k - x_k|.$$

The validity of Axioms 1–3 is obvious.

9. The following principle enables us to write down an infinite number of further examples: if $R = (X, \rho)$ is a metric space and M is an arbitrary subset of X, then M with the same function $\rho(x, y)$, but now assumed to be defined only for x and y in M, likewise forms a metric space; it is called a subspace of the space R.

(1) In the definition of a metric space we could have limited ourselves to two axioms for $\rho(x, y)$, namely:

1) $$\rho(x, y) = 0$$

if, and only if, $x = y$;

2) $$\rho(x, y) \le \rho(z, x) + \rho(z, y)$$

for arbitrary x, y, z.

It follows that

3) $$\rho(x, y) \ge 0,$$

4) $$\rho(x, y) = \rho(y, x)$$

and consequently Axiom 2 can be written in the form

2') $$\rho(x, y) \le \rho(x, z) + \rho(z, y).$$

(2) The set D^n of ordered n-tuples of real numbers with distance

$$\rho_p(x, y) = \left(\sum_{k=1}^n |y_k - x_k|^p\right)^{1/p} \qquad (p \ge 1)$$

also forms a metric space which we shall denote by $R_p{}^n$. Here the validity of Axioms 1 and 2 is again obvious. We shall check Axiom 3. Let

$$x = (x_1, x_2, \cdots, x_n), \qquad y = (y_1, y_2, \cdots, y_n) \quad \text{and} \quad z = (z_1, z_2, \cdots, z_n)$$

be points in $R_p{}^n$. If, as in Example 3, we set

$$y_k - x_k = a_k, \qquad z_k - y_k = b_k,$$

then the inequality

$$\rho_p(x, z) \leq \rho_p(x, y) + \rho_p(y, z)$$

assumes the form

$$(7) \qquad \left(\sum_{k=1}^{n} |a_k + b_k|^p\right)^{1/p} \leq \left(\sum_{k=1}^{n} |a_k|^p\right)^{1/p} + \left(\sum_{k=1}^{n} |b_k|^p\right)^{1/p}.$$

This is the so-called Minkowski inequality. Minkowski's inequality is obvious for $p = 1$ (since the absolute value of a sum is less than or equal to the sum of the absolute values) and therefore we can restrict ourselves to considering the case $p > 1$.

In order to prove inequality (7) for $p > 1$ we shall first establish Hölder's inequality:

$$(8) \qquad \sum_{k=1}^{n} |x_k y_k| \leq \left(\sum_{k=1}^{n} |x_k|^p\right)^{1/p} \left(\sum_{k=1}^{n} |y_k|^q\right)^{1/q},$$

where the number q is defined by the condition

$$(9) \qquad 1/p + 1/q = 1.$$

We note that inequality (8) is homogeneous in the sense that if it is satisfied for any two vectors

$$x = (x_1, x_2, \cdots, x_n) \quad \text{and} \quad y = (y_1, y_2, \cdots, y_n),$$

then it is also satisfied for the vectors λx and μy where λ and μ are arbitrary numbers. Therefore it is sufficient to prove inequality (8) for the case when

$$(10) \qquad \sum_{k=1}^{n} |x_k|^p = \sum_{k=1}^{n} |y_k|^q = 1.$$

Thus, we must prove that if Condition (10) is satisfied, then

$$(11) \qquad \sum_{k=1}^{n} |x_k y_k| \leq 1.$$

Consider in the (ξ, η)-plane the curve defined by the equation $\eta = \xi^{p-1}$, or equivalently by the equation $\xi = \eta^{q-1}$ (see Fig. 7). It is clear from the figure that for an arbitrary choice of positive values for a and b we have $S_1 + S_2 \geq ab$. If we calculate the areas S_1 and S_2, we obtain

$$S_1 = \int_0^a \xi^{p-1} \, d\xi = a^p/p; \qquad S_2 = \int_0^b \eta^{q-1} \, d\eta = b^q/q.$$

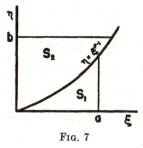

FIG. 7

Thus

$$ab \leq a^p/p + b^q/q.$$

Setting $a = |x_k|$, $b = |y_k|$ and summing with respect to k from 1 to n, we obtain

$$\sum_{k=1}^{n} |x_k y_k| \leq 1,$$

if we take (9) and (10) into consideration.

Inequality (11) and consequently the more general inequality (8) are thus proved. For $p = 2$ Hölder's inequality (8) becomes the Schwarz inequality (1).

We now proceed to the proof of the Minkowski inequality. Consider the identity

$$(|a| + |b|)^p = (|a| + |b|)^{p-1}|a| + (|a| + |b|)^{p-1}|b|.$$

Setting $a = x_k$, $b = y_k$ in the above identity and summing with respect to k from 1 to n, we obtain

$$\sum_{k=1}^{n} (|x_k| + |y_k|)^p = \sum_{k=1}^{n} (|x_k| + |y_k|)^{p-1}|x_k|$$
$$+ \sum_{k=1}^{n}(|x_k| + |y_k|)^{p-1}|y_k|.$$

If we now apply Hölder's inequality to each of the two sums on the right of the above equality and take into consideration the fact that $(p-1)q = p$, we obtain

$$\sum_{k=1}^{n} (|x_k| + |y_k|)^p$$
$$\leq \{\sum_{k=1}^{n} (|x_k| + |y_k|)^p\}^{1/q}\{(\sum_{k=1}^{n} |x_k|^p)^{1/p} + (\sum_{k=1}^{n} |y_k|^p)^{1/p}\}.$$

Dividing both sides of this inequality by

$$\{\sum_{k=1}^{n} (|x_k| + |y_k|)^p\}^{1/q},$$

we obtain

$$\{\sum_{k=1}^{n} (|x_k| + |y_k|)^p\}^{1/p} \leq (\sum_{k=1}^{n} |x_k|^p)^{1/p} + (\sum_{k=1}^{n} |y_k|^p)^{1/p},$$

whence inequality (7) follows immediately. This also establishes the triangle axiom for the space $R_p{}^n$.

(3) It is possible to show that the metric

$$\rho_0(x, y) = \max \{|y_k - x_k|; 1 \leq k \leq n\}$$

introduced in Example 4 can be defined in the following way:

$$\rho_0(x, y) = \lim_{p \to \infty} \left(\sum_{k=1}^{n} |y_k - x_k|^p\right)^{1/p}.$$

(4) From the inequality

$$ab \leq a^p/p + b^q/q \qquad\qquad (1/p + 1/q = 1)$$

established in Example (2) it is easy to deduce also the integral form of Hölder's inequality

$$\int_a^b x(t)y(t)\, dt \leq \left(\int_a^b |x^p(t)|\, dt\right)^{1/p} \left(\int_a^b |y^q(t)|\, dt\right)^{1/q},$$

which is valid for arbitrary functions $x(t)$ and $y(t)$ for which the integrals on the right have meaning. From this in turn we obtain the integral form of Minkowski's inequality:

$$\left(\int_a^b |x(t) + y(t)|^p\, dt\right)^{1/p} \leq \left(\int_a^b |x(t)|^p\, dt\right)^{1/p} + \left(\int_a^b |y(t)|^p\, dt\right)^{1/p}.$$

(5) We shall point out still another interesting example of a metric space. Its elements are all possible sequences of real numbers

$$x = (x_1, x_2, \cdots, x_n, \cdots)$$

such that $\sum_{k=1}^{\infty} |x_k|^p < \infty$, where $p \geq 1$ is any fixed number and the distance is defined by means of the formula

(12) $$\rho(x, y) = \left(\sum_{k=1}^{\infty} |y_k - x_k|^p\right)^{1/p}.$$

We shall denote this metric space by l_p.

By virtue of Minkowski's inequality (7) we have

$$\left(\sum_{k=1}^{n} |y_k - x_k|^p\right)^{1/p} \leq \left(\sum_{k=1}^{n} |x_k|^p\right)^{1/p} + \left(\sum_{k=1}^{n} |y_k|^p\right)^{1/p}$$

for arbitrary n. Since the series

$$\sum_{k=1}^{\infty} |x_k|^p \quad \text{and} \quad \sum_{k=1}^{\infty} |y_k|^p$$

converge by assumption, passing to the limit as $n \to \infty$ we obtain

(13) $$\left(\sum_{k=1}^{\infty} |y_k - x_k|^p\right)^{1/p} \leq \left(\sum_{k=1}^{\infty} |x_k|^p\right)^{1/p} + \left(\sum_{k=1}^{\infty} |y_k|^p\right)^{1/p},$$

and so the series on the left side also converges. This proves that formula (12), which defines distance in l_p, actually has meaning for arbitrary

$x, y \in l_p$. At the same time inequality (13) shows that the triangle axiom is satisfied in l_p. The remaining axioms are obvious.

§9. Convergence of sequences. Limit points

In §§9–11 we shall establish some fundamental concepts which we shall frequently use in the sequel.

An *open sphere* $S(x_0, r)$ in the metric space R is the set of all points $x \in R$ which satisfy the condition $\rho(x, x_0) < r$. The fixed point x_0 is called the *center* and the number r is called the *radius* of this sphere.

A *closed sphere* $S[x_0, r]$ is the set of all points $x \in R$ which satisfy the condition $\rho(x, x_0) \leq r$.

An ϵ-neighborhood of the point x, denoted by the symbol $O(x, \epsilon)$, is an open sphere of radius ϵ and center x_0.

A point x is called a *contact point* of the set M if every neighborhood of x contains at least one point of M. The set of all contact points of the set M is denoted by $[M]$ and is called the closure of M. Since every point belonging to M is obviously a contact point of M (each point is contained in every one of its neighborhoods), every set is contained in its closure: $M \subseteq [M]$.

THEOREM 1. *The closure of the closure of M is equal to the closure of M*:

$$[[M]] = [M].$$

Proof. Let $x \in [[M]]$. Then an arbitrary ϵ-neighborhood $O(x, \epsilon)$ of x contains a point $x_1 \in [M]$. Setting $\epsilon - \rho(x, x_1) = \epsilon_1$, we consider the sphere $O(x_1, \epsilon_1)$. This sphere lies entirely in the interior of $O(x, \epsilon)$. In fact, if $z \in O(x_1, \epsilon_1)$, then $\rho(z, x_1) < \epsilon_1$; and since $\rho(x, x_1) = \epsilon - \epsilon_1$, then, by the triangle axiom $\rho(z, x) \leq \epsilon_1 + (\epsilon - \epsilon_1) = \epsilon$, i.e. $z \in O(x, \epsilon)$. Since $x_1 \in [M]$, $O(x, \epsilon_1)$ contains a point $x_2 \in M$. But then $x_2 \in O(x, \epsilon)$. Since $O(x, \epsilon)$ is an arbitrary neighborhood of the point x, we have $x \in [M]$. This completes the proof of the theorem.

The validity of the following assertion is obvious.

THEOREM 2. *If $M_1 \subseteq M$, then $[M_1] \subseteq [M]$.*

THEOREM 3. *The closure of a sum is equal to the sum of the closures*:

$$[M_1 \cup M_2] = [M_1] \cup [M_2].$$

Proof. Let $x \in [M_1 \cup M_2]$, i.e. let an arbitrary neighborhood $O(x, \epsilon)$ contain the point $y \in M_1 \cup M_2$. If it were true that $x \notin [M_1]$ and $x \notin [M_2]$, we could find a neighborhood $O(x, \epsilon_1)$ which does not contain points of M_1 and a neighborhood $O(x, \epsilon_2)$ which does not contain points of M_2. But then the neighborhood $O(x, \epsilon)$, where $\epsilon = \min (\epsilon_1, \epsilon_2)$, would not contain points of $M_1 \cup M_2$. From the contradiction thus obtained it follows that x is contained in at least one of the sets $[M_1]$ and $[M_2]$, i.e.

$$[M_1 \cup M_2] \subseteq [M_1] \cup [M_2].$$

Since $M_1 \subseteq M_1 \cup M_2$ and $M_2 \subseteq M_1 \cup M_2$, the converse inclusion follows from Theorem 2.

The point x is called a *limit point* of the set M if an arbitrary neighborhood of x contains an infinite number of points of M.

A limit point of the set M can either belong to M or not. For example, if M is the set of rational numbers in the closed interval $[0, 1]$, then every point of this interval is a limit point of M.

A point x belonging to the set M is said to be an *isolated point* of this set if x has a neighborhood $O(x, \epsilon)$ which does not contain any points of M different from x.

THEOREM 4. *Every contact point of the set M is either a limit point of the set M or an isolated point of M.*

Proof. Let x be a contact point of the set M. This means that every neighborhood $O(x, \epsilon)$ of x contains at least one point belonging to M. Two cases are possible:

1) Every neighborhood of the point x contains an infinite number of points of the set M. In this case, x is a limit point of M.

2) We can find a neighborhood $O(x, \epsilon)$ of x which contains only a finite number of points of M. In this case, x will be an isolated point of the set M. In fact, let $x_1, x_2, \cdots, x_k$ be the points of M which are distinct from x and which are contained in the neighborhood $O(x, \epsilon)$. Further, let ϵ_0 be the least of the positive numbers $\rho(x, x_i)$, $i = 1, 2, \cdots, k$. Then the neighborhood $O(x, \epsilon_0)$ obviously does not contain any point of M distinct from x. The point x itself in this case must necessarily belong to M since otherwise $O(x, \epsilon_0)$ in general would not contain a single point of M, i.e. x would not be a contact point of the set M. This completes the proof of the theorem.

Thus, the set $[M]$ consists in general of points of three types:

1) Isolated points of the set M;

2) Limit points of the set M which belong to M;

3) Limit points of the set M which do not belong to M.

$[M]$ is obtained by adding to M all its limit points.

Let $x_1, x_2, \cdots$ be a sequence of points in the metric space R. We say that this sequence *converges to the point* x if every neighborhood $O(x, \epsilon)$ contains all points x_n starting with some one of them (i.e. if for every $\epsilon > 0$ we can find a natural number N_ϵ such that $O(x, \epsilon)$ contains all points x_n with $n > N_\epsilon$). The point x is said to be the *limit* of the sequence $\{x_n\}$.

This definition can obviously be formulated in the following form: the sequence $\{x_n\}$ converges to x if $\lim_{n\to\infty} \rho(x, x_n) = 0$.

The following assertions follow directly from the definition of limit: 1) no sequence can have two distinct limits; 2) if the sequence $\{x_n\}$ converges to the point x then every subsequence of $\{x_n\}$ converges to the same point x.

The following theorem establishes the close connection between the concepts of contact point and limit point on the one hand and the concept of limit on the other.

THEOREM 5. *A necessary and sufficient condition that the point x be a contact point of the set M is that there exist a sequence $\{x_n\}$ of points of the set M which converges to x; a necessary and sufficient condition that the point x be a limit point of M is that there exist a sequence of distinct points of the set M which converges to x.*

Proof. Necessity. If x is a contact point of the set M, then every neighborhood $O(x, 1/n)$ contains at least one point x_n of M. These points form a sequence which converges to x. If the point x is a limit point of M, every neighborhood $O(x, 1/n)$ contains a point $x_n \in M$ which is distinct from all the $x_i (i < n)$ (since the number of such points is finite). The points x_n are distinct and form a sequence which converges to x.

Sufficiency is obvious.

Let A and B be two sets in the metric space R. The set A is said to be *dense* in B if $[A] \supseteq B$. In particular, the set A is said to be *everywhere dense* in R if its closure $[A]$ coincides with the entire space R. For example, the set of rational numbers is everywhere dense on the real line.

EXAMPLES OF SPACES CONTAINING AN EVERYWHERE DENSE DENUMERABLE SET. (They are sometimes called "separable." For another definition of such spaces in terms of the concept of basis see §10, Theorem 4.) We shall consider the very same examples which were pointed out in §8.

1. The space described in Example 1, §8, is separable if, and only if, it consists of a denumerable number of points. This follows directly from the fact that in this space $[M] = M$ for an arbitrary set M.

All spaces enumerated in Examples 2–7, §8, are separable. We shall indicate a denumerable everywhere dense set in each of them and leave the details of the proof to the reader.

2. Rational points.

3. The set of all vectors with rational coordinates.

4. The set of all vectors with rational coordinates.

5. The set of all polynomials with rational coefficients.

6. The set of all sequences in each of which all terms are rational and only a finite (but arbitrary) number of terms is distinct from zero.

7. The set of all polynomials with rational coefficients.

The space of bounded sequences (Example 8, §8) is not separable. In fact, let us consider all possible sequences consisting of zeros and ones. They form a set with cardinal number that of the continuum (since each of them can be put into correspondence with the dyadic development of some real number which is contained in the interval [0, 1]). The distance between two such distinct points defined by formula (6), §8, is 1. We surround each of these points with a sphere of radius $\frac{1}{2}$. These spheres do not intersect. If some set is everywhere dense in the space under consideration, then each of the indicated spheres should contain at least one point of this set and consequently it cannot be denumerable.

(1) Let A be an arbitrary set in the metric space R and let x be a point in R. The distance from the point x to the set A is defined by the number

$$\rho(A, x) = \inf \{\rho(a, x); a \in A\}.$$

If $x \in A$, then $\rho(A, x) = 0$; but the fact that $\rho(A, x) = 0$ does not imply that $x \in A$. From the definition of contact point it follows immediately that $\rho(A, x) = 0$ if, and only if, x is a contact point of the set A.

Thus, the closure $[A]$ of the set A can be defined as the totality of all those points whose distance from the set A is zero.

(2) We can define the distance between two sets analogously. If A and B are two sets in R, then

$$\rho(A, B) = \inf \{\rho(a, b); a \in A, b \in B\}.$$

If $A \cap B \neq \theta$, then $\rho(A, B) = 0$; the converse is not true in general.

(3) If A is a set in the metric space R then the totality A' of its limit points is called its *derived set*.

Although the application to $[M]$ once more of the operation of closure always results again in $[M]$, the equality $(M')' = M'$ does not hold in general. In fact, if we take, for example, the set A of points of the form $1/n$ on the real line, then its derived set A' consists of the single point 0, but the set $A'' = (A')'$ will already be the void set. If we consider on the real line the set B of all points of the form $1/n + 1/(nm)$ ($n, m = 1, 2, \cdots$), then $B' = A \cup A'$, B'' is the point 0, and B''' is the void set.

§10. Open and closed sets

In this section we shall consider the more important types of sets in a metric space; these are the open and closed sets.

A set M in a metric space R is said to be *closed* if it coincides with its closure: $[M] = M$. In other words, a set is said to be closed if it contains all its limit points.

By Theorem 1, §9, the closure of an arbitrary set M is a closed set. Theorem 2, §9, implies that $[M]$ is the smallest closed set which contains M.

EXAMPLES. 1. An arbitrary closed interval $[a, b]$ on the real line is a closed set.

2. The closed sphere is a closed set. In particular, in the space $C[a, b]$ the set of functions satisfying the condition $|f| \leq K$ is closed.

3. The set of functions satisfying the condition $|f| < K$ (open sphere) is not closed; its closure is the set of functions satisfying the condition $|f| \leq K$.

4. Whatever the metric space R, the void set and the whole space R are closed sets.

5. Every set consisting of a finite number of points is closed.

The fundamental properties of closed sets can be formulated in the form of the following theorem.

THEOREM 1. *The intersection of an arbitrary number and the sum of an arbitrary finite number of closed sets are closed sets.*

Proof. Let $F = \bigcap_\alpha F_\alpha$, where the F_α are closed sets. Further, let x be a limit point of the set F. This means that an arbitrary neighborhood $O(x, \epsilon)$ of x contains an infinite number of points of F. But then $O(x, \epsilon)$ contains an infinite number of points of each F_α and consequently, since all the F_α are closed, the point x belongs to each F_α; thus, $x \in F = \bigcap_\alpha F_\alpha$, i.e. F is closed.

Now let F be the sum of a finite number of closed sets: $F = \bigcup_{i=1}^n F_i$, and let x be a point not belonging to F. We shall show that x cannot be a limit point of F. In fact, x does not belong to any of the closed sets F_i and consequently it is not a limit point of any of them. Therefore for every i we can find a neighborhood $O(x, \epsilon_i)$ of the point x which does not contain more than a finite number of points of F_i. If we take the smallest of the neighborhoods $O(x, \epsilon_1), \cdots, O(x, \epsilon_n)$, we obtain a neighborhood $O(x, \epsilon)$ of the point x which does not contain more than a finite number of points of F.

Thus, if the point x does not belong to F, it cannot be a limit point of F, i.e. F is closed. This completes the proof of the theorem.

The point x is said to be an *interior point* of the set M if there exists a neighborhood $O(x, \epsilon)$ of the point x which is contained entirely in M.

A set all of whose points are interior points is said to be an *open set*.

EXAMPLES. 6. The interval (a, b) of the real line D^1 is an open set; in fact, if $a < \alpha < b$, then $O(\alpha, \epsilon)$, where $\epsilon = \min (\alpha - a, b - \alpha)$, is contained entirely in the interval (a, b).

7. The open sphere $S(a, r)$ in an arbitrary metric space R is an open set. In fact, if $x \in S(a, r)$, then $\rho(a, x) < r$. We set $\epsilon = r - \rho(a, x)$. Then $S(x, \epsilon) \subseteq S(a, r)$.

8. The set of continuous functions satisfying the condition $|f| < K$, where K is an arbitrary number, is an open subset of the space $C[a, b]$.

THEOREM 2. *A necessary and sufficient condition that the set M be open is that its complement $R \setminus M$ with respect to the whole space R be closed.*

Proof. If M is open, then each point $x \in M$ has a neighborhood which belongs entirely to M, i.e. which does not have a single point in common with $R \setminus M$. Thus, no point which does not belong to $R \setminus M$ can be a contact point of $R \setminus M$, i.e. $R \setminus M$ is closed. Conversely, if $R \setminus M$ is closed, an arbitrary point of M has a neighborhood which lies entirely in M, i.e. M is open.

Since the void set and the whole space R are closed and are at the same time complements of each other, the theorem proved above implies the following corollary.

COROLLARY. The void set and whole space R are open sets.

The following important theorem which is the dual of Theorem 1 follows from Theorem 1 and the principle of duality established in §1 (the intersection of complements equals the complement of the sums, the sum of the complements equals the complement of the intersections).

THEOREM 1′. *The sum of an arbitrary number and the intersection of an arbitrary finite number of open sets are open sets.*

A family $\{G_\alpha\}$ of open sets in R is called a *basis* in R if every open set in R can be represented as the sum of a (finite or infinite) number of sets belonging to this family.

To check whether or not a given family of open sets is a basis we find the following criterion useful.

THEOREM 3. *A necessary and sufficient condition that a system of open sets $\{G_\alpha\}$ be a basis in R is that for every open set G and for every point $x \in G$ a set G_α can be found in this system such that $x \in G_\alpha \subset G$.*

Proof. If $\{G_\alpha\}$ is a basis, then every open set G is a sum of G_α's: $G = \bigcup_i G_{\alpha_i}$, and consequently every point x in G belongs to some G_α contained in G. Conversely, if the condition of the theorem is fulfilled, then $\{G_\alpha\}$ is a basis. In fact, let G be an arbitrary open set. For each point $x \in G$ we can find some $G_\alpha(x)$ such that $x \in G_\alpha \subset G$. The sum of these $G_\alpha(x)$ over all $x \in G$ equals G.

With the aid of this criterion it is easy to establish that in every metric space the family of all open spheres forms a basis. The family of all spheres with rational radii also forms a basis. On the real line a basis is formed, for example, by the family of all rational intervals (i.e. intervals with rational endpoints).

We shall say that a set is countable if it is either finite or denumerable.

R is said to be a *space with countable basis* or to satisfy the *second axiom of countability* if there is at least one basis in R consisting of a countable number of elements.

THEOREM 4. *A necessary and sufficient condition that R be a space with countable basis is that there exist in R an everywhere dense countable set.* (A finite everywhere dense set occurs only in spaces consisting of a finite set of points.)

Proof. Necessity. Let R have a countable basis $\{G_n\}$. Choose from each G_n an arbitrary point x_n. The set $\{x_n\}$ obtained in this manner is everywhere dense in R. In fact, let x be an arbitrary point in R and let $O(x, \epsilon)$ be a neighborhood of x. According to Theorem 3, a set G_n can be found such that $x \in G_n \subset O(x, \epsilon)$. Since G_n contains at least one of the points of the set $\{x_n\}$, any neighborhood $O(x, \epsilon)$ of an arbitrary point $x \in R$ contains at least one point from $\{x_n\}$ and this means that $\{x_n\}$ is everywhere dense in R.

Sufficiency. If $\{x_n\}$ is a countable everywhere dense set in R, then the family of spheres $S(x_n, 1/k)$ forms a countable basis in R. In fact, the set of all these spheres is countable (being the sum of a countable family of countable sets). Further, let G be an arbitrary open set and let x be any point in G. By the definition of an open set an $m > 0$ can be found such that the sphere $S(x, 1/m)$ lies entirely in G. We now select a point x_{n_0} from the set $\{x_n\}$ such that $\rho(x, x_{n_0}) < 1/3m$. Then the sphere $S(x_{n_0}, 1/2m)$ contains the point x and is contained in $S(x, 1/m)$ and consequently in G also. By virtue of Theorem 3 it follows from this that the spheres $S(x_n, 1/k)$ form a basis in R.

By virtue of this theorem, the examples introduced above (§ 9) of separable spaces are at the same time examples of spaces with countable basis.

We say that a system of sets M_α is a *covering* of the space R if $\bigcup M_\alpha = R$. A covering consisting of open (closed) sets will be called an *open (closed) covering*.

THEOREM 5. *If R is a metric space with countable basis, then we can select a countable covering from each of its open coverings.*

Proof. Let $\{O_\alpha\}$ be an arbitrary open covering of R. Thus, every point $x \in R$ is contained in some O_α.

Let $\{G_n\}$ be a countable basis in R. Then for every $x \in R$ there exists a $G_n(x) \in \{G_n\}$ and an α such that $x \in G_n(x) \subset O_\alpha$. The family of sets $G_n(x)$ selected in this way is countable and covers R. If we choose for each of the $G_n(x)$ one of the sets O_α containing it, we obtain a countable sub-covering of the covering $\{O_\alpha\}$.

It was already indicated above that the void set and the entire space R are simultaneously open and closed. A space in which there are no other sets which are simultaneously open and closed is said to be *connected*. The real line R^1 is one of the simplest examples of a connected metric space. But if we remove a finite set of points (for example, one point) from R^1, the remaining space is no longer connected. The simplest example of a space which is not connected is the space consisting of two points which are at an arbitrary distance from one another.

(1) Let M_k be the set of all functions f in $C[a, b]$ which satisfy a so-called Lipschitz condition

$$| f(t_1) - f(t_2) | \leq K | t_1 - t_2 |,$$

where K is a constant. The set M_k is closed. It coincides with the closure of the set of all differentiable functions which are such that $| f'(t) | \leq K$.

(2) The set $M = \bigcup_k M_k$ of all functions each of which satisfies a Lipschitz condition for some K is not closed. Since M contains the set of all polynomials, its closure is the entire space $C[a, b]$.

(3) Let distance be defined in the space X in two different ways, i.e. let there be given two distinct metrics $\rho_1(x, y)$ and $\rho_2(x, y)$. The metrics ρ_1 and ρ_2 are said to be *equivalent* if there exist two positive constants a and b such that $a < [\rho_1(x, y)/\rho_2(x, y)] < b$ for all $x \neq y$ in R. If an arbitrary set $M \subseteq X$ is closed (open) in the sense of the metric ρ_1, then it is closed (open) in the sense of an arbitrary metric ρ_2 which is equivalent to ρ_1.

(4) A number of important definitions and assertions concerning metric spaces (for example, the definition of connectedness) do not make use of the concept of metric itself but only of the concept of open (closed) set, or, what is essentially the same, the concept of neighborhood. In particular, in many questions the metric introduced in a metric space can be replaced by any other metric which is equivalent to the initial metric. This point of view leads naturally to the concept of topological space, which is a generalization of metric space.

A *topological space* is a set T of elements of an arbitrary nature (called points of this space) some subsets of which are labeled open sets. In this connection we assume that the following axioms are fulfilled:

1. T and the void set are open;

2. The sum of an arbitrary (finite or infinite) number and the intersection of an arbitrary finite number of open sets are open.

The sets $T \setminus G$, the complements of the open sets G with respect to T, are said to be closed. Axioms 1 and 2 imply the following two assertions.

1′. The void set and T are closed;

2′. The intersection of an arbitrary (finite or infinite) number and the sum of an arbitrary finite number of closed sets are closed.

A neighborhood of the point $x \in T$ is any open set containing x.

In a natural manner we introduce the concepts of contact point, limit point, and closure: $x \in T$ is said to be a contact point of the set M if every neighborhood of the point x contains at least one point of M; x is said to be a limit point of the set M if every neighborhood of the point x contains an infinite number of points of M. The totality of all contact points of the set M is called the closure $[M]$ of the set M.

It can easily be shown that closed sets (defined as the complements of open sets), and only closed sets, satisfy the condition $[M] = M$. As also in the case of a metric space $[M]$ is the smallest closed set containing M.

Similarly, as a metric space is the pair: set of points and a metric, so a topological space is the pair: set of points and a topology defined in this space. To introduce a topology into T means to indicate in T those subsets which are to be considered open in T.

EXAMPLES. (4-a) By virtue of Theorem 1′ open sets in a metric space satisfy Conditions 1 and 2 in the definition of a topological space. Thus, every metric space can be considered as a topological space.

(4-b) Let T consist of two points a and b and let the open sets in T be T, the void set, and the set consisting of the single point b. Axioms 1 and 2 are fulfilled. The closed sets are T, the void set, and the set consisting of the single point a. The closure of the set consisting of the point b is all of T.

(5) A topological space T is said to be *metrizable* if a metric can be introduced into the set T so that the sets which are open in the sense of this metric coincide with the open sets of the initial topological space. The space (4-b) is an example of a topological space which cannot be metrized.

(6) Although many fundamental concepts carry over from metric spaces to topological spaces defined in (4), this concept turns out to be too general in a number of cases. An important class of topological spaces consists of those spaces which satisfy, in addition to Axioms 1 and 2, the Hausdorff separation axiom:

3. Any two distinct points x and y of the space T have disjoint neighborhoods.

A topological space satisfying this axiom is called a *Hausdorff space*. Clearly, every metric space is a Hausdorff space. The space pointed out in Example (4-b) does not satisfy the Hausdorff axiom.

§11. Open and closed sets on the real line

The structure of open and closed sets in an arbitrary metric space can be very complicated. We shall now consider the simplest special case, namely that of open and closed sets on the real line. In this case their complete description does not present much of a problem and is given by the following theorem.

THEOREM 1. *Every open set on the real line is the sum of a countable number of disjoint intervals.*

Proof. [We shall also include sets of the form $(-\infty, \infty)$, (α, ∞), $(-\infty, \beta)$ as intervals.] Let G be an open set and let $x \in G$. Then by the definition of an open set we can find some interval I which contains the point x and belongs entirely to the set G. This interval can always be chosen so that its endpoints are rational. Having taken for every point $x \in G$ a corresponding interval I, we obtain a covering of the set G by means of a denumerable system of intervals (this system is denumerable because the set of all intervals with rational endpoints is denumerable). Furthermore, we shall say that the intervals I' and I'' (from the same covering) belong to one class if there exists a finite chain of intervals:

$$I' = I_1, I_2, \cdots, I_n = I''$$

(belonging to our covering) such that I_k intersects I_{k+1} ($1 \leq k \leq n - 1$). It is clear that there will be a countable number of such classes. Further, the union of all the intervals which belong to the same class obviously again forms an interval U of the same type, and intervals corresponding to distinct classes do not intersect. This completes the proof of the theorem.

Since closed sets are the complements of open sets, it follows that every closed set on the real line is obtained by removing a finite or denumerable number of open intervals on the real line.

The simplest examples of closed sets are segments, individual points, and the sum of a finite number of such sets. We shall now consider a more complicated example of a closed set on the real line, the so-called Cantor set.

Let F_0 be the closed interval [0, 1]. We remove the open interval $(\frac{1}{3}, \frac{2}{3})$ from F_0 and denote the remaining closed set by F_1. Then we remove the open intervals $(\frac{1}{9}, \frac{2}{9})$ and $(\frac{7}{9}, \frac{8}{9})$ from F_1 and denote the remaining closed set (consisting of four closed intervals) by F_2. From each of these four intervals we remove the middle interval of length $(\frac{1}{3})^3$, and so forth. If we continue this process, we obtain a decreasing sequence of closed sets F_n. We set $F = \bigcap_{n=0}^{\infty} F_n$; F is a closed set (since it is the intersection of the closed sets F_n). It is obtained from the closed interval [0, 1] by removing a denumerable number of open intervals. Let us consider the structure of the set F. The points

$$(1) \qquad 0, 1, \tfrac{1}{3}, \tfrac{2}{3}, \tfrac{1}{9}, \tfrac{2}{9}, \tfrac{7}{9}, \tfrac{8}{9}, \cdots$$

which are the endpoints of the deleted intervals obviously belong to F. However, the set F is not exhausted by these points. In fact, those points of the closed interval [0, 1] which belong to the set F can be characterized in the following manner. We shall write each of the numbers x, $0 \leq x \leq 1$, in the triadic system:

$$x = a_1/3 + a_2/3^2 + \cdots + a_n/3^n + \cdots ,$$

where the numbers a_n can assume the values 0, 1, and 2. As in the case of the ordinary decimal expansion, some numbers allow two different developments. For example,

$$\tfrac{1}{3} = \tfrac{1}{3} + \left(\tfrac{0}{3}\right)^2 + \cdots + \left(\tfrac{0}{3}\right)^n + \cdots = \tfrac{0}{3} + \left(\tfrac{2}{3}\right)^2 + \left(\tfrac{2}{3}\right)^3 + \cdots + \left(\tfrac{2}{3}\right)^n + \cdots .$$

It is easily verified that the set F contains those, and only those, numbers x, $0 \leq x \leq 1$, which can be written in at least one way in the form of a triadic fraction such that the number 1 does not appear in the sequence $a_1, a_2, \cdots, a_n, \cdots$. Thus, to each point $x \in F$ we can assign the sequence

$$(2) \qquad a_1, a_2, \cdots, a_n, \cdots ,$$

where a_n is 0 or 2. The set of all such sequences forms a set having the power of the continuum. We can convince ourselves of this by assigning to each sequence (2) the sequence

(2′) $$b_1, b_2, \cdots, b_n, \cdots,$$

where $b_n = 0$ if $a_n = 0$ and $b_n = 1$ if $a_n = 2$. The sequence (2′) can be considered as the development of a real number y, $0 \leq y \leq 1$, in the form of a dyadic fraction. We thus obtain a mapping of the set F onto the entire closed interval [0, 1]. This implies that F has the cardinal number of the continuum. [The correspondence established between F and the closed interval [0, 1] is single-valued but it is not one-to-one (because of the fact that the same number can sometimes be formed from distinct fractions). This implies that F has cardinal number not less than the cardinal number of the continuum. But F is a subset of the closed interval [0, 1] and consequently its cardinal number cannot be greater than that of the continuum. (See §5.)] Since the set of points (1) is denumerable, these points cannot exhaust all of F.

EXERCISE. Prove directly that the point $\frac{1}{4}$ belongs to the set F although it is not an endpoint of a single one of the intervals deleted. *Hint:* The point $\frac{1}{4}$ divides the closed interval [0, 1] in the ratio 1:3. The closed interval $[0, \frac{1}{3}]$ which remains after the first deletion is also divided in the ratio 1:3 by the point $\frac{1}{4}$, and so on.

The points (1) are said to be points of the first type of the set F and the remaining points are said to be points of the second type.

EXERCISE. Prove that the points of the first type form an everywhere dense set in F.

We have shown that the set F has the cardinal number of the continuum, i.e. that it contains as many points as the entire closed interval [0, 1].

It is interesting to compare this fact with the following result: the sum of the lengths of all the deleted intervals is $\frac{1}{3} + \frac{2}{9} + \frac{4}{27} + \cdots$, i.e. exactly 1!

§12. Continuous mappings. Homeomorphism. Isometry

Let $R = (X, \rho)$ and $R' = (Y, \rho')$ be two metric spaces. The mapping f of the space R into R' is said to be *continuous at the point* $x_0 \in R$ if for arbitrary $\epsilon > 0$ a $\delta > 0$ can be found such that

$$\rho'[f(x), f(x_0)] < \epsilon$$

for all x such that

$$\rho(x, x_0) < \delta.$$

In other words, the mapping f is continuous at the point x_0 if an arbitrary neighborhood $O(f(x_0), \epsilon)$ of the point $f(x_0)$ contains a neighborhood $O(x_0, \delta)$ of the point x_0 whose image is contained in the interior of $O(f(x_0), \epsilon)$.

A mapping f is said to be *continuous* if it is continuous at each point of the space R.

If R' is the real line, then a continuous mapping of R into R' is called a *continuous function* on R.

As in the case of the mapping of arbitrary sets we shall say that f is a mapping of R *onto* R' if every element $y \in R'$ has at least one inverse image.

In analysis, together with the definition of the continuity of a function "in terms of neighborhoods", the definition of continuity "in terms of sequences", which is equivalent to it, is widely used. The situation is analogous also in the case of continuous mappings of arbitrary metric spaces.

THEOREM 1. *A necessary and sufficient condition that the mapping f be continuous at the point x is that for every sequence $\{x_n\}$ which converges to x the corresponding sequence $\{f(x_n)\}$ converge to $y = f(x)$.*

Proof. The necessity is obvious. We shall prove the sufficiency of this condition. If the mapping f is not continuous at the point x, there exists a neighborhood $O(y, \epsilon)$ of the point $y = f(x)$ such that an arbitrary $O(x, \delta)$ contains points whose images do not belong to $O(y, \epsilon)$. Setting $\delta_n = 1/n$ $(n = 1, 2, \cdots)$, we select in each sphere $O(x, 1/n)$ a point x_n such that $f(x_n) \notin O(y, \epsilon)$. Then $x_n \to x$ but the sequence $\{f(x_n)\}$ does not converge to $f(x)$, i.e. the condition of the theorem is not satisfied, which was to be proved.

THEOREM 2. *A necessary and sufficient condition that the mapping f of the space R onto R' be continuous is that the inverse image of each closed set in R' be closed.*

Proof. Necessity. Let $M \subseteq R$ be the complete inverse image of the closed set $M' \subseteq R'$. We shall prove that M is closed. If $x \in [M]$, there exists a sequence $\{x_n\}$ of points in M which converges to x. But then, by Theorem 1, the sequence $\{f(x_n)\}$ converges to $f(x)$. Since $f(x_n) \in M'$ and M' is closed, we have $f(x) \in M'$; consequently $x \in M$, which was to be proved.

Sufficiency. Let x be an arbitrary point in R, $y = f(x)$, and let $O(y, \epsilon)$ be an arbitrary neighborhood of y. The set $R' \setminus O(y, \epsilon)$ is closed (since it is the complement of an open set). By assumption, $F = f^{-1}(R' \setminus O(y, \epsilon))$ is closed, and moreover, $x \notin F$. Thus, $R \setminus F$ is open and $x \in R \setminus F$; consequently, there is a neighborhood $O(x, \delta)$ of the point x which is contained in $R \setminus F$. If $z \in O(x, \delta)$, then $f(z) \in O(y, \epsilon)$, i.e. f is continuous, which was to be proved.

REMARK. The image of a closed set under a continuous mapping is not necessarily closed as is shown by the following example: map the half-open

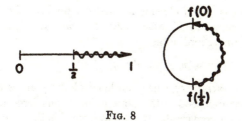

Fig. 8

interval $[0, 1)$ onto a circle of the same length. The set $[\frac{1}{2}, 1)$ which is closed in $[0, 1)$, goes over under this mapping into a set which is not closed (see Fig. 8).

Since in the case of a mapping "onto" the inverse image of the complement equals the complement of the inverse image, the following theorem which is the dual of Theorem 2 is valid.

THEOREM 2'. *A necessary and sufficient condition that the mapping f of the space R onto R' be continuous is that the inverse image of each open set in R' be open.*

The following theorem which is the analogue of the well-known theorem from analysis on the continuity of a composite function is valid for continuous mappings.

THEOREM 3. *If R, R', R'' are metric spaces and f and φ are continuous mappings of R into R' and R' into R'', respectively, then the mapping $z = \varphi(f(x))$ of the space R into R'' is continuous.*

The proof is carried out exactly as for real-valued functions.

The mapping f is said to be a *homeomorphism* if it is one-to-one and bicontinuous (i.e. both f and the inverse mapping f^{-1} are continuous).

The spaces R and R' are said to be *homeomorphic* if a homeomorphic correspondence can be established between them.

It is easy to see that two arbitrary intervals are homeomorphic, that an arbitrary open interval is homeomorphic to R^1, and so forth.

It follows from Theorems 2 and 2' of this section that a necessary and sufficient condition that a one-to-one mapping be a homeomorphism is that the closed (open) sets correspond to closed (open) sets.

This implies that a necessary and sufficient condition that a one-to-one mapping φ be a homeomorphism is that the equality

$$\varphi([M]) = [\varphi(M)]$$

hold for arbitrary M. (This follows from the fact that $[M]$ is the intersection of all closed sets which contain M, i.e. it is the minimal closed set which contains M.)

EXAMPLE. Consider the spaces $R^n = (D^n, \rho)$ and $R_0{}^n = (D^n, \rho_0)$ (see §8,

Examples 3 and 4). The following inequalities hold for the mapping which assigns to an element in R^n with coordinates $x_1, x_2, \ldots, x_n$ the element in R_0^n with the same coordinates:

$$\rho_0(x, y) \leq \rho(x, y) \leq n^{\frac{1}{2}}\rho_0(x, y).$$

Consequently an arbitrary ϵ-neighborhood of the point x of the space R^n contains a δ-neighborhood of the same point x considered as an element of the space R_0^n, and conversely. It follows from this that our mapping of R^n onto R_0^n is a homeomorphism.

An important special case of a homeomorphism is an isometric mapping.

We say that a one-to-one mapping $y = f(x)$ of a metric space R onto a metric space R' is *isometric* if

$$\rho(x_1, x_2) = \rho'[f(x_1), f(x_2)]$$

for arbitrary $x_1, x_2 \in R$. The spaces R and R' themselves, between which an isometric correspondence can be established, are said to be *isometric*.

The isometry of two spaces R and R' means that the metric relations between their elements are the same and that they can differ only in the nature of their elements, which is unessential. In the sequel we shall consider two isometric spaces simply as identical.

(1) The concept of continuity of a mapping can be defined not only for metric but also for arbitrary topological spaces. The mapping f of the topological space T into the topological space T' is said to be continuous at the point x_0 if for arbitrary neighborhood $O(y_0)$ of the point $y_0 = f(x_0)$ there exists a neighborhood $O(x_0)$ of the point x_0 such that $f(O(x_0)) \subset O(y_0)$.

Theorems 2 and 3 carry over automatically to continuous mappings of topological spaces.

§13. Complete metric spaces

From the very beginning of our study of mathematical analysis we are convinced of the important role in analysis that is played by the property of completeness of the real line, i.e. the fact that every fundamental sequence of real numbers converges to some limit. The real line represents the simplest example of the so-called complete metric spaces whose basic properties we shall consider in this section.

We shall call a sequence $\{x_n\}$ of points of a metric space R a *fundamental sequence* if it satisfies the Cauchy criterion, i.e. if *for arbitrary $\epsilon > 0$ there exists an N_ϵ such that $\rho(x_{n'}, x_{n''}) < \epsilon$ for all $n' \geq N_\epsilon, n'' \geq N_\epsilon$.*

It follows directly from the triangle axiom that every convergent sequence is fundamental. In fact, if $\{x_n\}$ converges to x, then for given $\epsilon > 0$ it is possible to find a natural number N_ϵ such that $\rho(x_n, x) < \epsilon/2$ for all $n \geq N_\epsilon$. Then $\rho(x_{n'}, x_{n''}) \leq \rho(x_{n'}, x) + \rho(x_{n''}, x) < \epsilon$ for arbitrary $n' \geq N_\epsilon$ and $n'' \geq N_\epsilon$.

DEFINITION 1. If every fundamental sequence in the space R converges to an element in R, R is said to be *complete*.

EXAMPLES. All the spaces considered in §8, with the exception of the one given in Example 7, are complete. In fact:

1. In the space consisting of isolated points (Example 1, §8) only those sequences in which there is a repetition of some point, beginning with some index, are fundamental. Clearly, every such sequence converges, i.e. this space is complete.

2. The completeness of the space R^1 of real numbers is known from analysis.

3. The completeness of the Euclidean space R^n follows directly from the completeness of R^1. In fact, let $\{x_i\}$ be a fundamental sequence; this means that for every $\epsilon > 0$ an $N = N_\epsilon$ can be found such that

$$\sum_{k=1}^n (x_p^{(k)} - x_q^{(k)})^2 < \epsilon^2$$

for all p, q greater than N. Then for each $k = 1, 2, \cdots, n$

$$| x_p^{(k)} - x_q^{(k)} | < \epsilon$$

for all $p, q > N$, i.e. $\{x_p^{(k)}\}$ is a fundamental sequence of real numbers. We set

$$x^{(k)} = \lim_{p\to\infty} x_p^{(k)},$$

and

$$x = (x^{(1)}, x^{(2)}, \cdots, x^{(n)}).$$

Then it is obvious that

$$\lim_{n\to\infty} x_n = x.$$

4. The completeness of the space R_0^n is proved in an exactly analogous manner.

5. We shall prove the completeness of the space $C[a, b]$. Let $\{x_n(t)\}$ be a fundamental sequence in $C[a, b]$. This means that for each $\epsilon > 0$ there exists an N such that $| x_n(t) - x_m(t) | < \epsilon$ for $n, m > N$ and all $t, a \le t \le b$. This implies that the sequence $\{x_n(t)\}$ converges uniformly and that its limit is a continuous function $x(t)$, where

$$| x_n(t) - x(t) | < \epsilon$$

for all t and for all n larger than some N; this means that $\{x_n(t)\}$ converges to $x(t)$ in the sense of the metric of the space $C[a, b]$.

6. The space l_2. Let $\{x^{(n)}\}$, where

$$x^{(n)} = (x_1^{(n)}, x_2^{(n)}, \cdots, x_k^{(n)}, \cdots),$$

be a fundamental sequence in l_2.

For arbitrary $\epsilon > 0$ an N can be found such that

(1) $\rho^2(x^{(n)}, x^{(m)}) = \sum_{k=1}^{\infty} (x_k^{(n)} - x_k^{(m)})^2 < \epsilon$ for $n, m > N$.

It follows from this that for arbitrary k

$$(x_k^{(n)} - x_k^{(m)})^2 < \epsilon,$$

i.e. for each k the sequence of real numbers $\{x_k^{(n)}\}$ converges. Set $\lim_{n\to\infty} x_k^{(n)} = x_k$. Denote the sequence $(x_1, x_2, \cdots, x_n, \cdots)$ by x. We must show that

a) $\sum_{k=1}^{\infty} x_k^2 < \infty$;   b) $\lim_{n\to\infty} \rho(x^{(n)}, x) = 0$.

To this end we shall write inequality (1) in the form

$$\sum_{k=1}^{\infty} (x_k^{(n)} - x_k^{(m)})^2$$

$$= \sum_{k=1}^{M} (x_k^{(n)} - x_k^{(m)})^2 + \sum_{k=M+1}^{\infty} (x_k^{(n)} - x_k^{(m)})^2 < \epsilon$$

(M arbitrary). Since each of these two sums is nonnegative, each of them is less than ϵ. Consequently

$$\sum_{k=1}^{M} (x_k^{(n)} - x_k^{(m)})^2 < \epsilon.$$

If we fix m in this inequality and pass to the limit as $n \to \infty$, we obtain

$$\sum_{k=1}^{M} (x_k - x_k^{(m)})^2 \leq \epsilon.$$

Since this inequality is valid for arbitrary M, we can pass to the limit as $M \to \infty$. We then obtain

$$\sum_{k=1}^{\infty} (x_k - x_k^{(m)})^2 \leq \epsilon.$$

The inequality thus obtained and the convergence of the series $\sum_{k=1}^{\infty} x_k^{(m)2}$ imply that the series $\sum_{k=1}^{\infty} x_k^2$ converges; consequently x is an element in l_2. Further, since ϵ is arbitrarily small, this inequality means that

$$\lim_{m\to\infty} \rho(x^{(m)}, x) = \lim_{m\to\infty} \{ \sum_{k=1}^{\infty} (x_k - x_k^{(m)})^2 \}^{\frac{1}{2}} = 0,$$

i.e. $x^{(n)} \to x$.

7. It is easy to convince ourselves of the fact that the space $C^2[a, b]$ is not complete. For example, the sequence of continuous functions

$$\varphi_n(t) = \arctan nt \qquad (-1 \leq t \leq 1)$$

is fundamental, but it does not converge to any continuous function (it converges in the sense of mean square deviation to the discontinuous function which is equal to $-\pi/2$ for $t < 0$, $\pi/2$ for $t > 0$, and 0 for $t = 0$).

EXERCISE. Prove that the space of all bounded sequences (Example 8, §8) is complete.

In analysis the so-called lemma on nested segments is widely used. In the theory of metric spaces an analogous role is played by the following theorem which is called the *principle of nested spheres*.

THEOREM 1. *A necessary and sufficient condition that the metric space R be complete is that every sequence of closed nested spheres in R with radii tending to zero have nonvoid intersection.*

Proof. Necessity. Assume the space R is complete and let S_1, S_2, S_3, $\cdots$ be a sequence of closed nested spheres. Let d_n be the diameter of the sphere S_n. By hypothesis $\lim_{n\to\infty} d_n = 0$. Denote the center of the sphere S_n by x_n. The sequence $\{x_n\}$ is fundamental. In fact, if $m > n$, then obviously $\rho(x_n, x_m) < d_n$. Since R is complete, $\lim_{n\to\infty} x_n$ exists. If we set

$$x = \lim_{n\to\infty} x_n,$$

then $x \in \bigcap_n S_n$. In fact, the sphere S_n contains all the points of the given sequence with the exception perhaps of the points $x_1, x_2, \cdots, x_{n-1}$. Thus, x is a limit point of each sphere S_n. But since S_n is a closed set, we have that $x \in S_n$ for all n.

Sufficiency. To prove the sufficiency we shall show that if the space R is not complete, i.e. if there exists a fundamental sequence in R which does not have a limit, then it is possible to construct a sequence of closed nested spheres in R whose diameters tend to zero and whose intersection is void. Let $\{x_n\}$ be a fundamental sequence of points in R which does not have a limit. We shall construct a sequence of closed spheres S_n in the following way. Let n_1 be such that $\rho(x_{n_1}, x_m) < \frac{1}{2}$ for all $m > n_1$. Denote by S_1 the sphere of radius 1 and center at x_{n_1}. Further, let $n_2 > n_1$ be such that $\rho(x_{n_2}, x_m) < \frac{1}{4}$ for all $m > n_2$. Denote by S_2 the sphere of radius $\frac{1}{2}$ with center x_{n_2}. Since by assumption $\rho(x_{n_1}, x_{n_2}) < \frac{1}{2}$, we have $S_2 \subset S_1$. Now let $n_3 > n_2$ be such that $\rho(x_{n_3}, x_m) < \frac{1}{8}$ for all $m > n_3$ and let S_3 be a sphere of radius $\frac{1}{4}$ with center x_{n_3}, and so forth. If we continue this construction we obtain a sequence of closed nested spheres $\{S_n\}$, where S_n has radius $(\frac{1}{2})^{n-1}$. This sequence of spheres has void intersection; in fact, if $x \in \bigcap_k S_k$, then $x = \lim_{n\to\infty} x_n$. As a matter of fact the sphere S_k contains all points x_n beginning with x_{n_k} and consequently $\rho(x, x_n) < (\frac{1}{2})^{k-1}$ for all $n > n_k$. But by assumption the sequence $\{x_n\}$ does not have a limit. Therefore $\bigcap S_n = \theta$.

If the space R is not complete, it is always possible to embed it in an entirely definite manner in a complete space.

DEFINITION 2. Let R be an arbitrary metric space. A complete metric space R^* is said to be the *completion* of the space R if: 1) R is a subspace of the space R^*; and 2) R is everywhere dense in R^*, i.e. $[R] = R^*$. (Here $[R]$ naturally denotes the closure of the space R in R^*.)

For example, the space of all real numbers is the completion of the space of rationals.

THEOREM 2. *Every metric space has a completion and all of its completions are isometric.*

Proof. We begin by proving uniqueness. It is necessary to prove that if R^* and R^{**} are two completions of the space R, then they are isometric, i.e. there is a one-to-one mapping φ of the space R^* onto R^{**} such that 1) $\varphi(x) = x$ for all $x \in R$; and 2) if $x^* \leftrightarrow x^{**}$ and $y^* \leftrightarrow y^{**}$, then $\rho(x^*, y^*) = \rho(x^{**}, y^{**})$.

Such a mapping φ is defined in the following way. Let x^* be an arbitrary point of R^*. Then, by the definition of completion, there exists a sequence $\{x_n\}$ of points in R which converges to x^*. But the sequence $\{x_n\}$ can be assumed to belong also to R^{**}. Since R^{**} is complete, $\{x_n\}$ converges in R^{**} to some point x^{**}. We set $\varphi(x^*) = x^{**}$. It is clear that this correspondence is one-to-one and does not depend on the choice of the sequence $\{x_n\}$ which converges to the point x^*. This is then the isometric mapping sought. In fact, by construction we have $\varphi(x) = x$ for all $x \in R$. Furthermore, if we let

$$\{x_n\} \rightarrow x^* \text{ in } R^* \text{ and } \{x_n\} \rightarrow x^{**} \text{ in } R^{**},$$

$$\{y_n\} \rightarrow y^* \text{ in } R^* \text{ and } \{y_n\} \rightarrow y^{**} \text{ in } R^{**},$$

then

$$\rho(x^*, y^*) = \lim_{n\to\infty} \rho(x_n, y_n)$$

and at the same time

$$\rho(x^{**}, y^{**}) = \lim_{n\to\infty} \rho(x_n, y_n).$$

Consequently

$$\rho(x^*, y^*) = \rho(x^{**}, y^{**}).$$

We shall now prove the existence of the completion. The idea involved in the proof is the same as that in the so-called Cantor theory of real numbers. The situation here is essentially even simpler than in the theory of real numbers since there it is required further that one define all the arithmetic operations for the newly introduced objects—the irrational numbers.

Let R be an arbitrary metric space. We shall say that two fundamental sequences $\{x_n\}$ and $\{x_n'\}$ in R are equivalent (denoting this by $\{x_n\} \sim \{x_n'\}$) if $\lim_{n\to\infty} \rho(x_n, x_n') = 0$. This equivalence relation is reflexive, symmetric and transitive. It follows from this that all fundamental sequences which can be constructed from points of the space R are partitioned into equivalence classes of sequences. We shall now define the space R^* in the following manner. The points of R^* will be all possible equivalence classes of fundamental sequences and the distance between points in R^* will be defined in the following way. Let x^* and y^* be two such classes. We choose one representative from each of these two classes, i.e. we select some fundamental sequence $\{x_n\}$ and $\{y_n\}$ from each, respectively. We set

(2) $$\rho(x^*, y^*) = \lim_{n\to\infty} \rho(x_n, y_n).$$

We shall prove the correctness of this definition of distance, i.e. we shall show that the limit (2) exists and does not depend on the choice of the representatives $\{x_n\} \in x^*$ and $\{y_n\} \in y^*$.

Since the sequences $\{x_n\}$ and $\{y_n\}$ are fundamental, with the aid of the triangle axiom we have for all sufficiently large n', n'':

$$| \rho(x_{n'}, y_{n'}) - \rho(x_{n''}, y_{n''}) |$$
$$= | \rho(x_{n'}, y_{n'}) - \rho(x_{n'}, y_{n''}) + \rho(x_{n'}, y_{n''}) - \rho(x_{n''}, y_{n''}) |$$
$$\leq | \rho(x_{n'}, y_{n'}) - \rho(x_{n'}, y_{n''}) | + | \rho(x_{n'}, y_{n''}) - \rho(x_{n''}, y_{n''}) |$$
$$\leq \rho(y_{n'}, y_{n''}) + \rho(x_{n'}, x_{n''}) < \epsilon/2 + \epsilon/2 = \epsilon.$$

Thus, the sequence of real numbers $s_n = \rho(x_n, y_n)$ satisfies the Cauchy criterion and consequently it has a limit. It remains to prove that this limit does not depend on the choice of $\{x_n\} \in x^*$ and $\{y_n\} \in y^*$. Let

$$\{x_n\}, \{x_n'\} \in x^* \quad \text{and} \quad \{y_n\}, \{y_n'\} \in y^*.$$

Now

$$\{x_n\} \sim \{x_n'\} \quad \text{and} \quad \{y_n\} \sim \{y_n'\}$$

imply that

$$| \rho(x_n, y_n) - \rho(x_n', y_n')|$$
$$= | \rho(x_n, y_n) - \rho(x_n', y_n) + \rho(x_n', y_n) - \rho(x_n', y_n')|$$
$$\leq | \rho(x_n, y_n) - \rho(x_n', y_n)| + | \rho(x_n', y_n) - \rho(x_n', y_n')|$$
$$\leq \rho(x_n, x_n') + \rho(y_n, y_n') \to 0,$$

i.e.

$$\lim_{n\to\infty} \rho(x_n, y_n) = \lim_{n\to\infty} \rho(x_n', y_n').$$

We shall now show that the metric space axioms are fulfilled in R^*.

Axiom 1 follows directly from the definition of equivalence of fundamental sequences.

Axiom 2 is obvious.

We shall now verify the triangle axiom. Since the triangle axiom is satisfied in the initial space R, we have

$$\rho(x_n, z_n) \leq \rho(x_n, y_n) + \rho(y_n, z_n).$$

Passing to the limit as $n \to \infty$, we obtain

$$\lim_{n\to\infty} \rho(x_n, z_n) \leq \lim_{n\to\infty} \rho(x_n, y_n) + \lim_{n\to\infty} \rho(y_n, z_n),$$

i.e.

$$\rho(x, z) \leq \rho(x, y) + \rho(y, z).$$

We shall now prove that R^* is the completion of the space R. (We use *the* keeping in mind that all completions of the space R are isometric.)

To each point $x \in R$ there corresponds some equivalence class of fundamental sequences, namely the totality of all sequences which converge to the point x.

We have:
if

$$x = \lim_{n\to\infty} x_n \text{ and } y = \lim_{n\to\infty} y_n,$$

then

$$\rho(x, y) = \lim_{n\to\infty} \rho(x_n, y_n).$$

Consequently, letting the corresponding class of fundamental sequences converging to x correspond to each point x we embed R isometrically in the space R^*.

In the sequel we shall not have to distinguish between the space R itself and its image in R^* (i.e. the totality of all equivalence classes of convergent sequences) and we can consider R to be a subset of R^*.

We shall now show that R is everywhere dense in R^*. In fact, let x^* be a point in R^* and let $\epsilon > 0$ be arbitrary. We select a representative in x^*, i.e. we choose a fundamental sequence $\{x_n\}$. Let N be such that $\rho(x_n, x_m) < \epsilon$ for all $n, m > N$. Then we have

$$\rho(x_n, x^*) = \lim_{m\to\infty} \rho(x_n, x_m) \leq \epsilon,$$

i.e. an arbitrary neighborhood of the point x^* contains a point of R. Thus we have $[R] = R^*$.

It remains to be proved that the space R^* is complete. We note, first of all, that by the construction of R^* an arbitrary fundamental sequence

$$(3) \qquad\qquad x_1, x_2, \cdots, x_n, \cdots$$

consisting of points belonging to R, converges in R^* to some point, namely to the point $x^* \in R^*$, defined by the sequence (3). Further, since R is dense in R^*, then for an arbitrary fundamental sequence $x_1^*, x_2^*, \cdots, x_n^*, \cdots$ of points in R^* we can construct an equivalent sequence $x_1, x_2, \cdots, x_n, \cdots$ consisting of points belonging to R. To do this it is sufficient to take for x_n any point in R such that $\rho(x_n, x_n^*) < 1/n$.

The sequence $\{x_n\}$ thus constructed will be fundamental and by what was proved above it will converge to some point $x^* \in R^*$. But then the sequence $\{x_n^*\}$ also converges to x^*. This proves the theorem completely.

§14. Principle of contraction mappings and its applications

As examples of the applications of the concept of completeness we shall consider the so-called contraction mappings which form a useful technique for the proof of various existence and uniqueness theorems (for example, in the theory of differential equations).

Let R be an arbitrary metric space. A mapping A of the space R into itself is said to be a *contraction* if there exists a number $\alpha < 1$ such that

$$(1) \qquad \rho(Ax, Ay) \leq \alpha\rho(x, y)$$

for any two points $x, y \in R$. Every contraction mapping is continuous. In fact, if $x_n \to x$, then, by virtue of (1), we also have $Ax_n \to Ax$.

THEOREM (PRINCIPLE OF CONTRACTION MAPPINGS). *Every contraction mapping defined in a complete metric space R has one and only one fixed point (i.e. the equation $Ax = x$ has one and only one solution).*

Proof. Let x_0 be an arbitrary point in R. Set $x_1 = Ax_0$, $x_2 = Ax_1 = A^2x_0$, and in general let $x_n = Ax_{n-1} = A^n x_0$. We shall show that the sequence $\{x_n\}$ is fundamental. In fact,

$$\rho(x_n, x_m) = \rho(A^n x_0, A^m x_0) \leq \alpha^n \rho(x_0, x_{m-n})$$

$$\leq \alpha^n \{\rho(x_0, x_1) + \rho(x_1, x_2) + \cdots + \rho(x_{m-n-1}, x_{m-n})\}$$

$$\leq \alpha^n \rho(x_0, x_1)\{1 + \alpha + \alpha^2 + \cdots + \alpha^{m-n-1}\} \leq \alpha^n \rho(x_0, x_1)\{1/(1 - \alpha)\}.$$

Since $\alpha < 1$, this quantity is arbitrarily small for sufficiently large n. Since R is complete, $\lim_{n \to \infty} x_n$ exists. We set $x = \lim_{n \to \infty} x_n$. Then by virtue of the continuity of the mapping A, $Ax = A \lim_{n \to \infty} x_n = \lim_{n \to \infty} Ax_n = \lim_{n \to \infty} x_{n+1} = x$.

Thus, the existence of a fixed point is proved. We shall now prove its uniqueness. If $Ax = x$, $Ay = y$, then $\rho(x, y) \leq \alpha\rho(x, y)$, where $\alpha < 1$; this implies that $\rho(x, y) = 0$, i.e. $x = y$.

The principle of contraction mappings can be applied to the proof of the existence and uniqueness of solutions obtained by the method of successive approximations. We shall consider the following simple examples.

1. $y = f(x)$, where $f(x)$ is a function defined on the closed interval $[a, b]$ satisfying the Lipschitz condition

$$|f(x_2) - f(x_1)| \leq K |x_2 - x_1|,$$

with $K < 1$, and mapping the closed interval $[a, b]$ into itself. Then f is a contraction mapping and according to the theorem proved above the

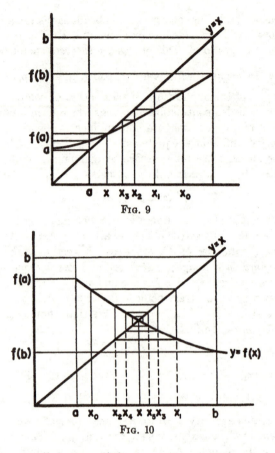

FIG. 9

FIG. 10

sequence x_0, $x_1 = f(x_0)$, $x_2 = f(x_1)$, $\cdots$ converges to the single root of the equation $x = f(x)$.

In particular, the condition of contraction is fulfilled if $|f'(x)| \leq K < 1$ on the closed interval $[a, b]$.

As an illustration, Figs. 9 and 10 indicate the course of the successive approximations in the case $0 < f'(x) < 1$ and in the case $-1 < f'(x) < 0$.

In the case where we are dealing with an equation of the form $F(x) = 0$, where $F(a) < 0$, $F(b) > 0$ and $0 < K_1 \leq F'(x) \leq K_2$ on $[a, b]$, a widely used method for finding its root amounts to setting $f(x) = x - \lambda F(x)$ and seeking a solution of the equation $x = f(x)$, which is equivalent to $F(x) = 0$. In fact, since $f'(x) = 1 - \lambda F'(x)$, $1 - \lambda K_2 \leq f'(x) \leq 1 - \lambda K_1$ and it is not difficult to choose λ so that we can apply the method of successive approximations.

2. Let us consider the mapping $y = Ax$ of the space R^n into itself given by the system of linear equations

$$y_i = \sum_{j=1}^{n} a_{ij}x_j + b_i \qquad (i = 1, 2, \cdots, n).$$

If Ax is a contraction mapping, we can apply the method of successive approximations to the solution of the equation $x = Ax$.

Under what conditions then is the mapping A a contraction? The answer to this question depends on the choice of the metric in R^n. (It is easy to see that with the metric (b) R^n is a metric space.)

a) $$\rho(x, y) = \max \{|x_i - y_i|; 1 \leq i \leq n\};$$

$$\rho(y', y'') = \max_i |y_i' - y_i''| = \max_i |\sum_j a_{ij}(x_j' - x_j'')|$$

$$\leq \max_i \sum_j |a_{ij}||x_j' - x_j''| \leq \max_i \sum_j |a_{ij}| \max_j |x_j' - x_j''|$$

$$= \max_i \sum_j |a_{ij}| \rho(x', x'').$$

This yields

$$(2) \qquad \sum_{j=1}^{n} |a_{ij}| \leq \alpha < 1$$

as the condition of contraction.

b) $$\rho(x, y) = \sum_{i=1}^{n} |x_i - y_i|;$$

$$\rho(y', y'') = \sum_i |y_i' - y_i''| = \sum_i |\sum_j a_{ij}(x_j' - x_j'')|$$

$$\leq \sum_i \sum_j |a_{ij}||x_j' - x_j''| \leq \max_j \sum_i |a_{ij}| \rho(x', x'').$$

This yields the following condition of contraction:

$$(3) \qquad \sum_i |a_{ij}| \leq \alpha < 1.$$

c) $$\rho(x, y) = \{\sum_{i=1}^{n} (x_i - y_i)^2\}^{\frac{1}{2}}.$$

Here

$$\rho^2(y', y'') = \sum_i \{\sum_j a_{ij}(x_j' - x_j'')\}^2 \leq \sum_i \sum_j a_{ij}^2 \rho^2(x', x'')$$

on the basis of the Schwarz inequality.

Then

$$(4) \qquad \sum_i \sum_j a_{ij}^2 \leq \alpha < 1$$

is the contraction condition.

Thus, in the case where one of the Conditions (2)–(4) is fulfilled there exists one and only one point $x = (x_1, x_2, \cdots, x_n)$ such that

$$x_i = \sum_{j=1}^{n} a_{ij}x_j + b_i,$$

where the successive approximations to this solution have the form:

$$x^{(0)} = (x_1^{\,0}, x_2^{\,0}, \cdots, x_n^{\,0});$$

$$x^{(1)} = (x_1^{\,1}, x_2^{\,1}, \cdots, x_n^{\,1});$$

$$\cdots\cdots\cdots\cdots\cdots\cdots\cdots\cdots$$

$$x^{(k)} = (x_1^{\,k}, x_2^{\,k}, \cdots, x_n^{\,k});$$

$$\cdots\cdots\cdots\cdots\cdots\cdots\cdots\cdots,$$

where

$$x_i^{\,k} = \sum_{j=1}^{n} a_{ij} x_j^{\,k-1} + b_i.$$

(Consequently any one of the Conditions (2)–(4) implies that

$$\begin{vmatrix} a_{11} - 1 & a_{12} & \cdots & a_{1n} \\ a_{21} & a_{22} - 1 & \cdots & a_{2n} \\ \cdots\cdots\cdots\cdots\cdots\cdots\cdots\cdots\cdots\cdots \\ a_{n1} & a_{n2} & \cdots & a_{nn} - 1 \end{vmatrix} \neq 0.)$$

Each of the Conditions (2)–(4) is *sufficient* in order that the mapping $y = Ax$ be a contraction. As concerns Condition (2) it could have been proved that it is also necessary in order that the mapping $y = Ax$ be a contraction (in the sense of the metric a)).

None of the Conditions (2)–(4) is necessary for the application of the method of successive approximations. Examples can be constructed in which any one of these conditions is fulfilled but the other two are not.

If $|a_{ij}| < 1/n$ (in this case all three conditions are fulfilled), then the method of successive approximations is applicable.

If $|a_{ij}| = 1/n$ (in this case all three sums equal 1), it is easy to see that the method of successive approximations is not applicable.

§15. Applications of the principle of contraction mappings in analysis

In the preceding section there were given some of the simplest examples of the application of the principle of contraction mappings in one- and n-dimensional spaces. However, the most essential applications for analysis of the principle of contraction mappings are in infinite-dimensional function spaces. Further on we shall show how with the aid of this principle we can obtain theorems on the existence and uniqueness of solutions for some types of differential and integral equations.

I. Let

$$(1) \qquad\qquad dy/dx = f(x, y)$$

be a given differential equation with the initial condition

$$(2) \qquad\qquad y(x_0) = y_0,$$

where $f(x, y)$ is defined and continuous in some plane region G which contains the point (x_0, y_0) and satisfies a Lipschitz condition with respect to y:

$$| f(x, y_1) - f(x, y_2)| \leq M | y_1 - y_2 |.$$

We shall prove that then on some closed interval $| x - x_0 | < d$ there exists a unique solution $y = \varphi(x)$ of the equation (1) satisfying the initial condition (2) (Picard's theorem).

Equation (1) together with the initial condition (2) is equivalent to the integral equation

$$(3) \qquad \varphi(x) = y_0 + \int_{x_0}^{x} f(t, \varphi(t))\, dt.$$

Since the function $f(x, y)$ is continuous, we have $| f(x, y)| \leq k$ in some region $G' \subseteq G$ which contains the point (x_0, y_0). Now we select a $d > 0$ such that the following conditions are fulfilled:

1) $(x, y) \in G'$ if $| x - x_0 | \leq d, | y - y_0 | \leq kd$;

2) $Md < 1$.

Denote by C^* the space of continuous functions φ^* which are defined on the closed interval $| x - x_0 | \leq d$ and are such that $| \varphi^*(x) - y_0 | \leq kd$ with the metric $\rho(\varphi_1, \varphi_2) = \max_x | \varphi_1(x) - \varphi_2(x)|$.

It is easy to see that C^* is a complete space. (This follows, for instance, from the fact that a closed subset of a complete space is a complete space.) Let us consider the mapping $\psi = A\varphi$ defined by the formula

$$\psi(x) = y_0 + \int_{x_0}^{x} f(t, \varphi(t))\, dt,$$

where $| x - x_0 | \leq d$. This is a contraction mapping of the complete space C^* into itself. In fact, let $\varphi \in C^*, | x - x_0 | \leq d$. Then

$$| \psi(x) - y_0 | = \left| \int_{x_0}^{x} f(t, \varphi(t))\, dt \right| \leq kd$$

and consequently $A(C^*) \subseteq C^*$. Moreover, we have

$$| \psi_1(x) - \psi_2(x) | \leq \int_{x_0}^{x} | f(t, \varphi_1(t)) - f(t, \varphi_2(t)) |\, dt$$

$$\leq Md \max_x | \varphi_1(x) - \varphi_2(x) |.$$

Since $Md < 1$, the mapping A is a contraction.

From this it follows that the operator equation $\varphi = A\varphi$ (and consequently equation (3) also) has one and only one solution.

II. Let

(4) $$\varphi_i'(x) = f_i(x_0, \varphi_1(x), \cdots, \varphi_n(x)); \quad i = 1, 2, \cdots, n;$$

be a given system of differential equations with initial conditions

(5) $$\varphi_i(x_0) = y_{0i}; \quad i = 1, 2, \cdots, n;$$

where the functions $f_i(x, y_1, \cdots, y_n)$ are defined and continuous in some region G of the space R^{n+1} such that G contains the point $(x_0, y_{01}, \cdots, y_{0n})$ and satisfy a Lipschitz condition

$$|f_i(x, y_1^{(1)}, \cdots, y_n^{(1)}) - f_i(x, y_1^{(2)}, \cdots, y_n^{(2)})|$$
$$\leq M \max\{|y_i^{(1)} - y_i^{(2)}|; 1 \leq i \leq n\}.$$

We shall prove that then on some closed interval $|x - x_0| < d$ there exists one and only one system of solutions $y_i = \varphi_i(x)$ satisfying system (4) and the initial conditions (5).

System (4) together with the initial conditions (5) is equivalent to the system of integral equations

(6) $$\varphi_i(x) = y_{0i} + \int_{x_0}^x f_i(t, \varphi_1(t), \cdots, \varphi_n(t)) \, dt; \quad i = 1, \cdots, n.$$

Since the function f_i is continuous in some region $G' \subseteq G$ containing the point $(x_0, y_{01}, \cdots, y_{0n})$, the inequalities $|f_i(x, y_1, \cdots, y_n)| \leq K$, where K is a constant, are fulfilled.

We now choose $d > 0$ to satisfy the following conditions:

1) $(x, y_1, \cdots, y_n) \in G$ if $|x - x_0| \leq d$, $|y_i - y_{0i}| \leq Kd$,

2) $Md < 1$.

We now consider the space C_n^* whose elements are ordered systems $\bar{\varphi} = (\varphi_1(x), \cdots, \varphi_n(x))$ consisting of n functions which are defined and continuous for all x for which $|x - x_0| \leq d$ and such that $|\varphi_i(x) - y_{0i}| \leq Kd$, with the metric

$$\rho(\bar{\varphi}, \bar{\psi}) = \max_{x, i} |\varphi_i(x) - \psi_i(x)|.$$

The mapping $\bar{\psi} = A\bar{\varphi}$, given by the system of integral equations

$$\psi_i(x) = y_{0i} + \int_{x_0}^x f(t, \varphi_1(t), \cdots, \varphi_n(t)) \, dt,$$

is a contraction mapping of the complete space C_n^* into itself. In fact,

$$\psi_i^{(1)}(x) - \psi_i^{(2)}(x) = \int_{x_0}^x [f(t, \varphi_1^{(1)}, \cdots, \varphi_n^{(1)}) - f(t, \varphi_1^{(2)}, \cdots, \varphi_n^{(2)})] \, dt$$

and consequently

$$\max_{x, i} |\psi_i^{(1)}(x) - \psi_i^{(2)}(x)| \leq Md \max_{x, i} |\varphi_i^{(1)}(x) - \varphi_i^{(2)}(x)|.$$

Since $Md < 1$, A is a contraction mapping.

It follows that the operator equation $\bar{\varphi} = A\bar{\varphi}$ has one and only one solution.

III. We shall now apply the method of contraction mappings to the proof of the existence and uniqueness of the solution of the Fredholm nonhomogeneous linear integral equation of the second kind:

$$(7) \qquad f(x) = \lambda \int_a^b K(x, y)f(y) \, dy + \varphi(x),$$

where $K(x, y)$ (the so-called *kernel*) and $\varphi(x)$ are given functions, $f(x)$ is the function sought, and λ is an arbitrary parameter.

We shall see that our method is applicable only in the case of sufficiently small values of the parameter λ.

We shall assume that $K(x, y)$ and $\varphi(x)$ are continuous for $a \le x \le b$; $a \le y \le b$ and consequently that $|K(x, y)| \le M$. Consider the mapping $g = Af$, i.e. $g(x) = \lambda \int_a^b K(x, y)f(y) \, dy + \varphi(x)$, of the complete space $C[a, b]$ into itself. We obtain

$$\rho(g_1, g_2) = \max |g_1(x) - g_2(x)| \le |\lambda| M(b - a) \max |f_1 - f_2|.$$

Consequently, the mapping A is a contraction for $|\lambda| < 1/M(b - a)$.

From this, on the basis of the principle of contraction mappings, we can conclude that the Fredholm equation has a unique continuous solution for every $|\lambda| < 1/M(b - a)$. The successive approximations to this solution: $f_0(x), f_1(x), \cdots, f_n(x), \cdots$ have the form

$$f_n(x) = \lambda \int_a^b K(x, y)f_{n-1}(y) \, dy + \varphi(x).$$

IV. This method is applicable also in the case of nonlinear equations of the form

$$(8) \qquad f(x) = \lambda \int_a^b K(x, y, f(y)) \, dy + \varphi(x),$$

where K and φ are continuous. Furthermore K satisfies the condition

$$|K(x, y, z_1) - K(x, y, z_2)| \le M |z_1 - z_2|$$

for $|\lambda| < 1/M(b - a)$ since here again for the mapping $g = Af$ of the complete space $C[a, b]$ into itself given by the formula

$$g(x) = \lambda \int_a^b K(x, y, f(y)) \, dy + \varphi(x)$$

the inequality

$$\max \mid g_1(x) - g_2(x) \mid \leq \mid \lambda \mid M (b - a) \max \mid f_1 - f_2 \mid$$

holds.

V. Consider the Volterra type integral equation

$$(9) \qquad f(x) = \lambda \int_a^x K(x, y) f(y) \, dy + \varphi(x)$$

which differs from an equation of Fredholm type in that the upper limit in the integral is the variable quantity x. This equation can be considered as a particular case of the Fredholm equation if we complete the definition of the function $K(x, y)$ for $y > x$ by means of the equation $K(x, y) = 0$ (for $y > x$).

In contrast to the Fredholm integral equation for which we were required to limit ourselves to small values of the parameter λ the principle of contraction mappings (and the method of successive approximations based on it) is applicable to Volterra equations for all values of the parameter λ. We note first of all that the principle of contraction mappings can be generalized in the following manner: *if A is a continuous mapping of a complete metric space R into itself such that the mapping A^n is a contraction for some n, then the equation*

$$Ax = x$$

has one and only one solution.

In fact, if we take an arbitrary point $x \in R$ and consider the sequence $A^{kn}x$ ($k = 0, 1, 2, \cdots$), a repetition of the argument introduced in §14 yields the convergence of this sequence. Let $x_0 = \lim_{k \to \infty} A^{kn}x$. Then $Ax_0 = x_0$. In fact, $Ax_0 = \lim_{k \to \infty} A^{kn}Ax$. Since the mapping A^n is a contraction, we have

$$\rho(A^{kn}Ax, A^{kn}x) \leq \alpha\rho(A^{(k-1)n}Ax, A^{(k-1)n}x) \leq \cdots \leq \alpha^k\rho(Ax, x).$$

Consequently,

$$\lim_{k \to \infty} \rho(A^{kn}Ax, A^{kn}x) = 0,$$

i.e. $Ax_0 = x_0$.

Now consider the mapping

$$g(x) = \lambda \int_a^x K(x, y) f(y) \, dy + \varphi(x) = Af.$$

If f_1 and f_2 are two continuous functions defined on the closed interval $[a, b]$, then

$$\mid Af_1(x) - Af_2(x) \mid = \left| \lambda \int_a^x K(x, y)[f_1(y) - f_2(y)] \, dy \right| \leq \lambda \, Mm(x - a),$$

$$(M = \max | K(x, y)|, \qquad m = \max (| f_1 - f_2 |)),$$

$$| A^2 f_1(x) - A^2 f_2(x) | \leq \lambda^2 Mm(x - a)^2/2, \cdots,$$

$$| A^n f_1(x) - A^n f_2(x)| \leq \lambda^n Mm(x - a)^n/n! \leq \lambda^n Mm(b - a)^n/n!$$

For an arbitrary value of λ the number n can be chosen so large that

$$\lambda^n(b - a)^n/n! < 1,$$

i.e. the mapping A^n will be a contraction. Consequently, the Volterra equation (9) has a solution for arbitrary λ and this solution is unique.

§16. Compact sets in metric spaces

A set M in the metric space R is said to be *compact* if every sequence of elements in M contains a subsequence which converges to some $x \in R$.

Thus, for example, by virtue of the Bolzano-Weierstrass theorem every bounded set on the real line is compact. Other examples of compact sets will be given below. It is clear that an arbitrary subset of a compact set is compact.

The concept of total boundedness which we shall now introduce is closely related to the concept of compactness.

Let M be any set in the metric space R and let ϵ be a positive number. The set A in R is said to be an ϵ-net with respect to M if for an arbitrary point $x \in M$ at least one point $a \in A$ can be found such that

$$\rho(a, x) < \epsilon.$$

For example, the lattice points form a $2^{\frac{1}{2}}$-net in the plane. A subset M of R is said to be *totally bounded* if R contains a finite ϵ-net with respect to M for every $\epsilon > 0$. It is clear that a totally bounded set is bounded since if an ϵ-net A can be found for M consisting of a finite number of points, then A is bounded and since the diameter of M does not exceed diameter $A + 2\epsilon$, M is also bounded; as Example 2 below will show, the converse is not true in general.

The following obvious remark is often useful: if the set M is totally bounded, then its closure $[M]$ is totally bounded.

It follows at once from the definition of total boundedness that every totally bounded metric space R with an infinite number of points is separable. In fact, construct a finite $(1/n)$-net in R for every n. Their sum over all n is a denumerable set which is everywhere dense in R.

EXAMPLES. 1. For subsets of Euclidean n-space total boundedness coincides with ordinary boundedness, i.e. with the possibility of enclosing a given set in the interior of some sufficiently large cube. In fact, if such a cube is subdivided into cubicles with diagonal of length $\epsilon/n^{\frac{1}{2}}$, then the vertices of these cubicles will form a finite ϵ-net in the initial cube and therefore also in an arbitrary set which lies in the interior of this cube.

2. The unit sphere S in the space l_2 is an example of a bounded set which is not totally bounded. In fact, consider in S points of the form

$$e_1 = (1, 0, 0, \cdots),$$
$$e_2 = (0, 1, 0, \cdots),$$
$$\cdots\cdots\cdots\cdots\cdots$$

The distance between any two such points e_n and e_m $(m \neq n)$ is $2^{\frac{1}{2}}$. From this it is clear that in S there can be no finite ϵ-net for any $\epsilon \leq 2^{-\frac{1}{2}}$.

3. Consider in l_2 the set Π of points $x = (x_1, x_2, \cdots, x_n, \cdots)$ which satisfy the following conditions:

$$|x_1| \leq 1, \qquad |x_2| \leq \tfrac{1}{2}, \cdots, \qquad |x_n| \leq (\tfrac{1}{2})^n, \cdots.$$

This set is called the fundamental parallelopiped in the space l_2. It is an example of an infinite-dimensional totally bounded set. To prove that this set is totally bounded we proceed as follows.

Let $\epsilon > 0$ be given. Choose n so that $(\tfrac{1}{2})^n < \epsilon/2$. Associate with each point

$$(1) \qquad\qquad x = (x_1, x_2, \cdots, x_n, \cdots)$$

in Π the point

$$(2) \qquad\qquad x^* = (x_1, x_2, \cdots, x_n, 0, 0, \cdots)$$

from the same set. Then we have

$$\rho(x, x^*) = \left(\sum_{k=n+1}^{\infty} x_k^2\right)^{\frac{1}{2}} \leq \left(\sum_{k=n+1}^{\infty} 1/4^k\right)^{\frac{1}{2}} < (\tfrac{1}{2})^n < \epsilon/2.$$

The subset Π^* of points of the form (2) of Π is totally bounded (since it is a bounded set in n-dimensional space). We choose a finite $(\epsilon/2)$-net in Π^*. It is clear that it will be at the same time an ϵ-net for Π.

The following theorem establishes the interrelation among the concepts of compactness, completeness, and total boundedness.

THEOREM 1. *A necessary and sufficient condition that a subset M of a complete metric space R be compact is that M be totally bounded.*

Proof. Necessity. We shall assume that M is not totally bounded, i.e. we shall assume that for some $\epsilon > 0$ a finite ϵ-net cannot be found in M. We take an arbitrary point x_1 in M. Since, by assumption, there is no finite ϵ-net in M, a point $x_2 \in M$ can be found such that $\rho(x_1, x_2) \geq \epsilon$. Further, there exists a point $x_3 \in M$ such that $\rho(x_1, x_3) \geq \epsilon$ and $\rho(x_2, x_3) \geq \epsilon$ (otherwise the points x_1 and x_2 would form an ϵ-net in M), and so on; continuing this process we construct a sequence $x_1, x_2, \cdots, x_n, \cdots$ of points in R which satisfy the condition $\rho(x_m, x_n) \geq \epsilon$ for $m \neq n$. It is clear that it is impossible to select any convergent subsequence from such a sequence.

Sufficiency. Let R be complete and let M be totally bounded. We shall prove that in M we can select a convergent subsequence from every sequence. Let $\{x_n\}$ be a sequence of points in M. Set

$$\epsilon_1 = 1, \ \epsilon_2 = \tfrac{1}{2}, \ \cdots, \ \epsilon_k = 1/k, \ \cdots$$

and construct for every ϵ_k a corresponding finite ϵ_k-net in M:

$$a_1{}^{(k)}, \ a_2{}^{(k)}, \ \cdots, \ a_{n_k}{}^{(k)}.$$

Describe about each of the points which form a 1-net in M a sphere of radius 1. Since these spheres cover all of M and are finite in number, at least one of them, let us call it S_1, contains an infinite subsequence $x_1{}^{(1)}$, $x_2{}^{(1)}, \ \cdots, \ x_n{}^{(1)}, \ \cdots$ of the sequence $\{x_n\}$. Further, about each of the points which form a $\tfrac{1}{2}$-net in R we describe a sphere of radius $\tfrac{1}{2}$. Since the number of these spheres is again finite, at least one of them, let us call it S_2, contains an infinite subsequence $x_1{}^{(2)}, \ x_2{}^{(2)}, \ \cdots, \ x_n{}^{(2)}, \ \cdots$ of the sequence $\{x_n{}^{(1)}\}$. Further, we find a sphere S_3 of radius $\tfrac{1}{3}$ containing an infinite subsequence $x_1{}^{(3)}, \ x_2{}^{(3)}, \ \cdots, \ x_n{}^{(3)}, \ \cdots$ of the sequence $\{x_n{}^{(2)}\}$, and so forth.

Now we choose from the sequences

$$x_1{}^{(1)}, \ x_2{}^{(1)}, \ \cdots, \ x_n{}^{(1)}, \ \cdots,$$

$$x_1{}^{(2)}, \ x_2{}^{(2)}, \ \cdots, \ x_n{}^{(2)}, \ \cdots,$$

$$\cdots \cdots \cdots \cdots \cdots \cdots \cdots \cdots \cdots$$

the "diagonal" subsequence

$$x_1{}^{(1)}, \ x_2{}^{(2)}, \ \cdots, \ x_n{}^{(n)}, \ \cdots;$$

this subsequence is fundamental because all its terms beginning with $x_n{}^{(n)}$ lie in the interior of the sphere S_n of radius $1/n$. Since the space R is complete, this subsequence has a limit. This completes the proof of the theorem.

The following generalization of Theorem 1 is often useful.

THEOREM 2. *A necessary and sufficient condition that a subset M of a complete metric space R be compact is that for every $\epsilon \geq 0$ there exist in R a compact ϵ-net for M.*

Proof. The necessity is trivial; we shall prove the sufficiency. Let $\epsilon > 0$ be given and let A be a compact $(\epsilon/2)$-net for M. We choose a finite $(\epsilon/2)$-net in A. It is clear that it will be a finite ϵ-net for M; consequently, by virtue of the preceding theorem, the set M is compact.

§17. Arzelà's theorem and its applications

The application of Theorems 1 and 2 of the preceding section, which yield necessary and sufficient conditions for compactness to individual special cases, is not always simple. For sets, situated in some given special space, special criteria for compactness can be given which are more suitable for practical application.

In analysis one of the more important metric spaces is the space $C[a, b]$.

For subsets of this space the most important and frequently employed criterion of compactness is given by Arzelà's theorem.

In order to formulate this theorem it is necessary to first introduce the following concepts.

A family $\{\varphi(x)\}$ of functions defined on a closed interval is said to be *uniformly bounded* if there exists a number M such that $|\varphi(x)| < M$ for all x and for all φ belonging to the given family.

A family of functions is said to be *equicontinuous* if for every $\epsilon > 0$ there is a $\delta > 0$ such that

$$|\varphi(x_1) - \varphi(x_2)| < \epsilon$$

for all x_1, x_2 such that $|x_1 - x_2| < \delta$ and for all φ in the given family.

ARZELÀ'S THEOREM. *A necessary and sufficient condition that a family of continuous functions defined on the closed interval $[a, b]$ be compact in $C[a, b]$ is that this family be uniformly bounded and equicontinuous.*

Proof. Necessity. Let the set Φ be compact in $C[a, b]$. Then, by Theorem 1, §16, for each positive ϵ there exists a finite $(\epsilon/3)$-net $\varphi_1, \varphi_2, \cdots, \varphi_k$ in Φ. Each of the functions φ_i, being a continuous function on a closed interval is bounded: $|\varphi_i| \leq M_i$.

Set $M = \max M_i + \epsilon/3$. By the definition of an $(\epsilon/3)$-net, for every $\varphi \in \Phi$ we have for at least one φ_i,

$$\rho(\varphi, \varphi_i) = \max |\varphi(x) - \varphi_i(x)| < \epsilon/3.$$

Consequently

$$|\varphi| < |\varphi_i| + \epsilon/3 < M_i + \epsilon/3 < M.$$

Thus, Φ is uniformly bounded.

Further, since each of the functions φ_i is continuous and consequently uniformly continuous on $[a, b]$, for a given $\epsilon/3$ there exists a δ_i such that

$$|\varphi_i(x_1) - \varphi_i(x_2)| < \epsilon/3$$

if $|x_1 - x_2| < \delta_i$.

Set $\delta = \min \delta_i$. Then for $|x_1 - x_2| < \delta$ and for any $\varphi \in \Phi$, taking φ_i so that $\rho(\varphi, \varphi_i) < \epsilon/3$, we have:

$|\varphi(x_1) - \varphi(x_2)|$
$$= |\varphi(x_1) - \varphi_i(x_1) + \varphi_i(x_1) - \varphi_i(x_2) + \varphi_i(x_2) - \varphi(x_2)|$$
$$\leq |\varphi(x_1) - \varphi_i(x_1)| + |\varphi_i(x_1) - \varphi_i(x_2)| + |\varphi_i(x_2) - \varphi(x_2)|$$
$$< \epsilon/3 + \epsilon/3 + \epsilon/3 = \epsilon.$$

The equicontinuity of Φ is thus proved.

Sufficiency. Let Φ be a uniformly bounded and equicontinuous family of functions. In accordance with Theorem 1, §16, in order to prove its compactness in $C[a, b]$ it is sufficient to show that for arbitrary $\epsilon > 0$ there exists for it in $C[a, b]$ a finite ϵ-net. Let

$$|\varphi| \leq M \text{ for all } \varphi \in \Phi$$

and let $\delta > 0$ be chosen so that

$$|\varphi(x_1) - \varphi(x_2)| < \epsilon/5 \quad \text{for} \quad |x_1 - x_2| < \delta; \quad x_1, x_2 \in [a, b];$$

and for all $\varphi \in \Phi$. Subdivide the segment $[a, b]$ on the x-axis by means of the points $x_0 = a, x_1, x_2, \cdots, x_n = b$ into intervals of length less than δ and construct vertical lines at these points of subdivision. We subdivide the segment $[-M, M]$ on the y-axis by means of the points $y_0 = -M$, $y_1, y_2, \cdots, y_m = M$ into intervals of length $\epsilon/5$ and construct horizontal lines at these points of subdivision. Thus we subdivide the rectangle $a \leq x \leq b, -M \leq y \leq M$ into cells with horizontal sides of length less than δ and vertical sides of length $\epsilon/5$. We now assign to every function $\varphi \in \Phi$ a polygonal arc $\psi(x)$ with vertices at the points (x_k, y_l), i.e. at vertices of the constructed net and deviating at the points x_k from the function φ by less than $\epsilon/5$ (the existence of such a polygonal arc is obvious).

Since, by construction,

$$|\varphi(x_k) - \psi(x_k)| < \epsilon/5,$$

$$|\varphi(x_{k+1}) - \psi(x_{k+1})| < \epsilon/5,$$

and

$$|\varphi(x_k) - \varphi(x_{k+1})| < \epsilon/5,$$

we have

$$|\psi(x_k) - \psi(x_{k+1})| < 3\epsilon/5.$$

Since the function $\psi(x)$ is linear between the points x_k and x_{k+1}, we have

$$|\psi(x_k) - \psi(x)| < 3\epsilon/5 \quad \text{for all } x, x_k \leq x \leq x_{k+1}.$$

Now let x be an arbitrary point of the closed interval $[a, b]$ and let x_k be the subdivision point chosen which is closest to x from the left. Then

$$|\varphi(x) - \psi(x)| \leq |\varphi(x) - \varphi(x_k)| + |\varphi(x_k) - \psi(x_k)| + |\psi(x_k) - \psi(x)| < \epsilon.$$

Consequently the polygonal arcs $\psi(x)$ form an ϵ-net with respect to Φ. Their number is finite (since only a finite number of polygonal arcs can be drawn through a finite number of points); thus, Φ is totally bounded. This proves the theorem completely.

Arzelà's theorem has many applications. We shall demonstrate its application in the example of the following existence theorem for ordinary differential equations with continuous right member.

THEOREM 2 (PEANO). *Let*

$$dy/dx = f(x, y)$$

be a given differential equation. If the function $f(x, y)$ is continuous in a closed region G, then at least one integral curve of the given equation passes through each interior point (x_0, y_0) of this region.

Proof. Since $f(x, y)$ is continuous in the closed region, it is bounded:

$$| f(x, y) | < M.$$

We draw straight lines with slopes M and $-M$ through the point (x_0, y_0). Further, we draw vertical lines $x = a$ and $x = b$ so that the two triangles cut off by them with the common vertex (x_0, y_0) lie entirely in the interior of the region G.

We now construct a so-called Euler polygonal arc for the given equation in the following way: we draw from the point (x_0, y_0) a straight line with the slope $f(x_0, y_0)$. On this straight line we take a point (x_1, y_1) and draw through it a straight line with the slope $f(x_1, y_1)$. On this straight line we take a point (x_2, y_2) and so forth. We now consider a sequence of Euler polygonal arcs $L_1, L_2, \cdots, L_k, \cdots$, passing through the point (x_0, y_0) such that the length of the greatest of the links of the polygonal arc L_k tends to zero as $k \to \infty$. Let $\varphi_k(x)$ be the function whose graph is the polygonal arc L_k. The functions $\varphi_1, \varphi_2, \cdots, \varphi_k, \cdots$ possess the following properties: they are 1) defined on the same segment $[a, b]$, 2) all bounded, and 3) equicontinuous.

On the basis of Arzelà's theorem, we can choose a uniformly convergent subsequence from the sequence $\{\varphi_k(x)\}$. Let this sequence be

$$\varphi^{(1)}(x), \varphi^{(2)}(x), \cdots, \varphi^{(k)}(x), \cdots .$$

We set $\varphi(x) = \lim_{k\to\infty} \varphi^{(k)}(x)$. It is clear that $\varphi(x_0) = y_0$. It remains to verify that $\varphi(x)$ satisfies the given differential equation on the closed interval $[a, b]$. To do this it is necessary to show that for arbitrary $\epsilon > 0$

$$\left| \frac{\varphi(x'') - \varphi(x')}{x'' - x'} - f(x', \varphi(x')) \right| < \epsilon$$

provided the quantity $| x'' - x' |$ is sufficiently small. To prove this it is sufficient in turn to establish that for sufficiently large k

$$\left| \frac{\varphi^{(k)}(x'') - \varphi^{(k)}(x')}{x'' - x'} - f(x', \varphi(x')) \right| < \epsilon$$

if only the difference $| x'' - x' |$ is sufficiently small.

Since $f(x, y)$ is continuous in the region G, for arbitrary $\epsilon > 0$ an $\eta > 0$ can be found such that

$$f(x', y') - \epsilon < f(x, y) < f(x', y') + \epsilon, \qquad (y' = \varphi(x')),$$

if

$$|x - x'| < 2\eta \quad \text{and} \quad |y - y'| < 4M\eta.$$

The set of points $(x, y) \in G$ which satisfy both these inequalities form some rectangle Q. Now let K be so large that for all $k > K$

$$|\varphi(x) - \varphi^{(k)}(x)| < M\eta$$

and all links of the polygonal arc L_k have length less than η. Then if $|x' - x| < 2\eta$, all the Euler polygons $y = \varphi^{(k)}(x)$ for which $k > K$ lie entirely in the interior of Q.

Further, let $(a_0, b_0), (a_1, b_1), \cdots, (a_{n+1}, b_{n+1})$ be vertices of the polygonal arc $y = \varphi^{(k)}(x)$, where

$$a_0 \leq x' < a_1 < a_2 < \cdots < a_n < x'' \leq a_{n+1}$$

(we assume $x'' > x'$ for definiteness; we can consider the case $x'' < x'$ analogously). Then

$$\varphi^{(k)}(a_1) - \varphi^{(k)}(x') = f(a_0, b_0)(a_1 - x');$$

$$\varphi^{(k)}(a_{i+1}) - \varphi^{(k)}(a_i) = f(a_i, b_i)(a_{i+1} - a_i), \qquad i = 1, 2, \cdots, n - 1;$$

$$\varphi^{(k)}(x'') - \varphi^{(k)}(a_n) = f(a_n, b_n)(x'' - a_n).$$

From this it follows that for $|x'' - x'| < \eta$,

$$[f(x', y') - \epsilon](a_1 - x') < \varphi^{(k)}(a_1) - \varphi^{(k)}(x') < [f(x', y') + \epsilon](a_1 - x');$$

$$[f(x', y') - \epsilon](a_{i+1} - a_i) < \varphi^{(k)}(a_{i+1}) - \varphi^{(k)}(a_i)$$

$$< [f(x', y') + \epsilon](a_{i+1} - a_i), \qquad i = 1, 2, \cdots, n - 1;$$

$$[f(x', y') - \epsilon](x'' - a_n) < \varphi^{(k)}(x'') - \varphi^{(k)}(a_n) < [f(x', y') + \epsilon](x'' - a_n).$$

Summing these inequalities, we obtain

$$[f(x', y') - \epsilon](x'' - x') < \varphi^{(k)}(x'') - \varphi^{(k)}(x') < [f(x', y') + \epsilon](x'' - x'),$$

which was to be proved.

The solution $\varphi(x)$ thus obtained will in general not be the unique solution of the equation $y' = f(x, y)$ which passes through the point (x_0, y_0).

§18. Compacta

In §16 we said that a subset M of a metric space R is compact if every sequence of elements in M has a subsequence which converges to some $x \in R$.

In this connection the limit point x might belong but it also might not belong to the set M. If from every sequence of elements in M it is possible to select a subsequence which converges to some x *belonging to M*, then the set M is said to be *compact in itself*. For the space R itself (as moreover also for all its closed subsets) the concepts of compactness and compactness in itself coincide. A compact metric space is called a *compactum* for brevity.

Inasmuch as compactness in itself of a set is an intrinsic property of this set which does not depend on the metric space in which it is embedded, it is natural to limit oneself to the study of compacta, i.e. to consider each such set simply as a separate metric space.

THEOREM 1. *A necessary and sufficient condition that a set M compact in a metric space R be a compactum is that M be closed in R.*

Proof. Necessity. Suppose M is not closed; then we can find in M a sequence $\{x_n\}$ which converges to a point $x \notin M$. But then this sequence cannot contain a subsequence which converges to some point $y \in M$, i.e. M cannot be a compactum.

Sufficiency. Since M is compact, every sequence $\{x_n\} \subset M$ contains a subsequence which converges to some $x \in R$. If M is closed, $x \in M$, i.e. M is compact in itself.

The next corollary follows from this and Theorem 1, §16.

COROLLARY. *Every closed bounded subset of Euclidean space is a compactum.*

THEOREM 2. *A necessary and sufficient condition that a metric space be a compactum is that it be: 1) complete and 2) totally bounded.*

The proof of this theorem is made by a verbatim repetition of the proof of Theorem 1, §16.

THEOREM 3. *Every compactum K contains a countable everywhere dense set.*

Proof. Since a compactum is a totally bounded space, K contains a finite $(1/n)$-net: $a_1, a_2, \cdots, a_{m_n}$ for every n. The union of all these ϵ-nets is a finite or denumerable set. This and Theorem 4, §10, imply the following corollary.

COROLLARY. *Every compactum has a countable basis.*

THEOREM 4. *A necessary and sufficient condition that the metric space R be a compactum is that either of the following two conditions holds:*

1) *An arbitrary open covering $\{O_\alpha\}$ of the space R contains a finite subcovering;*

2) *An arbitrary system $\{F_\alpha\}$ of closed sets in R with the finite intersection property has a nonvoid intersection.* (A system of sets is said to *have the finite intersection property* if an arbitrary finite number of these sets has nonvoid intersection.)

Proof. We note first of all that the equivalence of the two conditions formulated above follows directly from the principle of duality (§1). In fact, if $\{O_\alpha\}$ is an open covering of the space R, then $\{R \setminus O_\alpha\}$ is a system of closed sets satisfying the condition

$$\cap(R \setminus O_\alpha) = \theta.$$

The condition that we can select a finite subcovering from $\{O_\alpha\}$ is equivalent to the fact that the system of closed sets $\{R \setminus O_\alpha\}$ cannot have the finite intersection property if it has a void intersection.

We shall now prove that Condition 1 is necessary and sufficient that R be a compactum.

Necessity. Let R be a compactum and let $\{O_\alpha\}$ be an open covering of R. We choose in R for each $n = 1, 2, \cdots$ a finite $(1/n)$-net consisting of the points $a_k^{(n)}$ and we enclose each of these points with the sphere

$$S(a_k^{(n)}, 1/n)$$

of radius $1/n$. It is clear that for arbitrary n

$$R = \cup_k S(A_k^{(n)}, 1/n).$$

We shall assume that it is impossible to choose a finite system of sets covering K from $\{O_\alpha\}$. Then for each n we can find at least one sphere $S(a_{k(n)}^{(n)}, 1/n)$ which cannot be covered by a finite number of the sets O_α. We choose such a sphere for each n and consider the sequence of their centers $\{a_{k(n)}^{(n)}\}$. Since R is a compactum, there exists a point $\xi \in R$ which is the limit of a subsequence of this sequence. Let O_β be a set of $\{O_\alpha\}$ which contains ξ. Since O_β is open, we can find an $\epsilon > 0$ such that $S\{\xi, \epsilon\} \subset O_\beta$. We now choose an index n and a point $a_{k(n)}^{(n)}$ so that $\rho(\xi, a_{k(n)}^{(n)}) < \epsilon/2$, $1/n < \epsilon/2$. Then, obviously, $S(a_{k(n)}^{(n)}, 1/n) \subset O_\beta$, i.e. the sphere $S(a_{k(n)}^{(n)}, 1/n)$ is covered by a single set O_β. The contradiction thus obtained proves our assertion.

Sufficiency. We assume that the space R is such that from each of its open coverings it is possible to select a finite subcovering. We shall prove that R is a compactum. To do this it is sufficient to prove that R is complete and totally bounded. Let $\epsilon > 0$. Take a neighborhood $O(x, \epsilon)$ about each of the points $x \in R$; we then obtain an open covering of R. We choose from this covering a finite subcovering $O(x_1, \epsilon), \cdots, O(x_n, \epsilon)$. It is clear that the centers $x_1, \cdots, x_n$ of these neighborhoods form a finite ϵ-net in R. Since $\epsilon > 0$ is arbitrary, it follows that R is totally bounded. Now let $\{S_n\}$ be a sequence of nested closed spheres whose radii tend to zero. If their intersection is void, then the sets $R \setminus S_n$ form an open covering of R from which it is impossible to select a finite subcovering. Thus, from Condition 1 it follows that R is complete and totally bounded, i.e. that R is compact.

THEOREM 5. *The continuous image of a compactum is a compactum.*

Proof. Let Y be a compactum and let $Y = \varphi(X)$ be its continuous image. Further, let $\{O_\alpha\}$ be an open covering of the space Y. Set $U_\alpha = \varphi^{-1}(O_\alpha)$.

Since the inverse image of an open set under a continuous mapping is open, it follows that $\{U_\alpha\}$ is an open covering of the space X. Since X is a compactum, we can select a finite subcovering $U_1, U_2, \cdots, U_n$ from the covering $\{U_\alpha\}$. Then the corresponding sets $O_1, O_2, \cdots, O_n$ form a finite subcovering of the covering $\{O_\alpha\}$.

THEOREM 6. *A one-to-one mapping of a compactum which is continuous in one direction is a homeomorphism.*

Proof. Let φ be a one-to-one continuous mapping of the compactum X onto the compactum Y. Since, according to the preceding theorem, the continuous image of a compactum is a compactum, the set $\varphi(M)$ is a compactum for an arbitrary closed $M \subset X$ and consequently $\varphi(M)$ is closed in Y. It follows that the inverse image under the mapping φ^{-1} of an arbitrary closed set $M \subset X$ is closed; i.e. the mapping φ^{-1} is continuous.

REMARK. The following result follows from Theorem 6: if the equation

$$(3) \qquad\qquad dy/dx = f(x, y),$$

where the function $f(x, y)$ is continuous in a closed bounded region G which contains the point (x_0, y_0) for every y_0 belonging to some closed interval $[a, b]$, has a unique solution satisfying the initial condition $y(x_0) = y_0$, then this solution is a continuous function of the initial value y_0.

In fact, since the function $f(x, y)$ is continuous in a closed bounded region, it is bounded and consequently the set P of solutions of equation (3) corresponding to initial values belonging to the closed interval $[a, b]$ is uniformly bounded and equicontinuous. Moreover, the set P is closed. In fact, if $\{\varphi_n(x)\}$ is a sequence of solutions of equation (3) which converges uniformly to a function $\varphi(x)$, then $\varphi(x)$ is also a solution of equation (3) since, if

$$\varphi_n' = f(x, \varphi_n(x)),$$

then passing to the limit as $n \to \infty$, we obtain

$$\varphi' = f(x, \varphi(x)).$$

We have

$$\varphi(x_0) = \lim \varphi_n(x_0) \in [a, b].$$

In virtue of Arzelà's theorem and Theorem 1 of this section it follows from this that P is a compactum.

We set the point $\varphi(x_0)$ of the closed interval $[a, b]$ into correspondence with each solution $\varphi(x)$ of equation (3). By assumption this correspondence is one-to-one. Moreover, since

$$\max | \varphi^{(1)}(x) - \varphi^{(2)}(x) | \geq | \varphi^{(1)}(x_0) - \varphi^{(2)}(x_0) |,$$

the mapping $\varphi(x) \to \varphi(x_0)$ is continuous. By virtue of Theorem 6 the inverse mapping is also continuous and this then signifies the continuous dependence of the solutions on the initial conditions.

Now let C_{XY} be the set of all continuous mappings $y = f(x)$ of a compactum X into a compactum Y. We introduce distance into C_{XY} by means of the formula

$$\rho(f, g) = \sup \{\rho[f(x), g(x)]; \quad x \in X\}.$$

It is easy to verify that in this way C_{XY} is transformed into a metric space.

THEOREM 7 (GENERALIZED THEOREM OF ARZELÀ). *A necessary and sufficient condition that a set $D \subset C_{XY}$ be compact in C_{XY} is that in D the functions $y = f(x)$ be equicontinuous, i.e. that for arbitrary $\epsilon > 0$ there exist a $\delta > 0$ such that*

(1) $$\rho(x', x'') < \delta$$

implies

(2) $$\rho[f(x'), f(x'')] < \epsilon$$

for arbitrary f in D and x', x'' in X.

Proof. We embed C_{XY} in the space M_{XY} of all mappings of the compactum X into the compactum Y with the same metric

$$\rho(f, g) = \sup \{\rho[f(x), g(x)]; \quad x \in X\}$$

which was introduced in C_{XY} and prove the compactness of the set D in M_{XY}. Since C_{XY} is closed in M_{XY}, the compactness of the set D in M_{XY} will imply its compactness in C_{XY}. (That C_{XY} is closed in M_{XY} follows from the fact that the limit of a uniformly convergent sequence of continuous mappings is also a continuous mapping. The indicated assertion is a direct generalization of the known theorem in analysis and is proved in exactly the same way.)

Let $\epsilon > 0$ be arbitrary and choose δ such that (1) implies (2) for all f in D and all x', x'' in X. Let the points $x_1, x_2, \cdots, x_n$ form a $(\delta/2)$-net in X. It is easy to see that X can be represented as the sum of nonintersecting sets ϵ_i such that $x, y \in \epsilon_i$ implies that $\rho(x, y) < \delta$. In fact, it is sufficient to set, for example,

$$\epsilon_i = S(x_i, \delta/2) \setminus \bigcup_{j < i} S(x_j, \delta/2).$$

We now consider in the compactum Y a finite ϵ-net $y_1, y_2, \cdots, y_m$; we denote the totality of functions $g(x)$ which assume the values y_j on the sets ϵ_i by L. The number of such functions is clearly finite. We shall show that they form a 2ϵ-net in M_{XY} with respect to D. In fact, let $f \in D$. For every point x_i in x_1, $\cdots$, x_n we can find a point y_j in y_1, $\cdots$, y_m such that

$$\rho[f(x_i), y_j] < \epsilon.$$

Let the function $g(x) \in L$ be chosen so that $g(x_i) = y_j$. Then

$$\rho[f(x), g(x)] \leq \rho[f(x), f(x_i)] + \rho[f(x_i), g(x_i)] + \rho[g(x), g(x_i)] < 2\epsilon$$

if i is chosen so that $x \in \epsilon_i$.

From this it follows that $\rho(f, g) < 2\epsilon$ and thus the compactness of D in M_{XY} and consequently in C_{XY} also is proved.

§19. Real functions in metric spaces

A real function on a space R is a mapping of R into the space R^1 (the real line).

Thus, for example, a mapping of R^n into R^1 is an ordinary real-valued function of n variables.

In the case when the space R itself consists of functions, the functions of the elements of R are usually called *functionals*. We introduce several examples of functionals of functions $f(x)$ defined on the closed interval $[0, 1]$:

$$F_1(f) = \sup f(x);$$

$$F_2(f) = \inf f(x);$$

$$F_3(f) = f(x_0) \quad \text{where} \quad x_0 \in [0, 1];$$

$$F_4(f) = \varphi[f(x_0), f(x_1), \cdots, f(x_n)] \quad \text{where} \quad x_i \in [0, 1]$$

and the function $\varphi(y_1, \cdots, y_n)$ is defined for all real y_i;

$$F_5(f) = \int_0^1 \varphi[x, f(x)] \, dx,$$

where $\varphi(x, y)$ is defined and continuous for all $0 \leq x \leq 1$ and all real y;

$$F_6(f) = f'(x_0);$$

$$F_7(f) = \int_0^1 [1 + f'^2(x)]^{\frac{1}{2}} \, dx;$$

$$F_8(f) = \int_0^1 | f'(x) | \, dx.$$

Functionals can be defined on all of R or on a subset of R. For example, in the space C the functionals F_1, F_2, F_3, F_4, F_5 are defined on the entire space, $F_6(f)$ is defined only for functions which are differentiable at the point x_0, $F_7(f)$ for functions for which $[1 + f'^2(x)]^{\frac{1}{2}}$ is integrable, and $F_8(f)$ for functions for which $| f'(x) |$ is integrable.

The definition of continuity for real functions and functionals remains the same as for mappings in general (see §12).

For example, $F_1(f)$ is a continuous functional in C since

$$\rho(f, g) = \sup |f - g| \quad \text{and} \quad |\sup f - \sup g| \leq \sup |f - g|;$$

F_2, F_3, F_5 are also continuous functionals in C; F_4 is continuous in the space C if the function φ is continuous for all arguments; F_6 is discontinuous at every point in the space C for which it is defined. In fact, let $g(x)$ be such that $g'(x_0) = 1, |g(x)| < \epsilon$ and $f = f_0 + g$. Then $f'(x_0) = f_0'(x_0) + 1$ and $\rho(f, f_0) < \epsilon$. This same functional is continuous in the space $C^{(1)}$ of functions having a continuous derivative with the metric

$$\rho(f, g) = \sup [|f - g| + |f' - g'|];$$

F_7 is also a discontinuous functional in the space C. In fact, let $f_0(x) \equiv 0$ and $f_n(x) = (1/n) \sin 2\pi nx$. Then $\rho(f_n, f_0) = 1/n \to 0$. However, $F_7(f_n)$ is a constant (it does not depend on n) which is greater than $(17)^{\frac{1}{2}}$ and $F_7(f_0) = 1$.

Consequently, $F_7(f)$ is discontinuous at the point f_0.

By virtue of this same example $F_8(f)$ is also discontinuous in the space C. Both functionals F_7 and F_8 are continuous in the space $C^{(1)}$.

The following theorems which are the generalizations of well-known theorems of elementary analysis are valid for real functions defined on compacta.

THEOREM 1. *A continuous real function defined on a compactum is uniformly continuous.*

Proof. Assume f is continuous but not uniformly continuous, i.e. assume there exist x_n and x_n' such that

$$|x_n - x_n'| < 1/n \quad \text{and} \quad |f(x_n) - f(x_n')| \geq \epsilon.$$

From the sequence $\{x_n\}$ we can choose a subsequence $\{x_{n_k}\}$ which converges to x. Then also $\{x_{n_k}'\} \to x$ and either $|f(x) - f(x_n')| \geq \epsilon/2$ or $|f(x) - f(x_n)| \geq \epsilon/2$, which contradicts the continuity of $f(x)$.

THEOREM 2. *If the function $f(x)$ is continuous on the compactum K, then f is bounded on K.*

Proof. If f were not bounded on K, then there would exist a sequence $\{x_n\}$ such that $f(x_n) \to \infty$. We choose from $\{x_n\}$ a subsequence which converges to x: $\{x_{n_k}\} \to x$. Then in an arbitrarily small neighborhood of x the function $f(x)$ will assume arbitrarily large values which contradicts the continuity of f.

THEOREM 3. *A function f which is continuous on a compactum K attains its least upper and greatest lower bounds on K.*

Proof. Let $A = \sup f(x)$. Then there exists a sequence $\{x_n\}$ such that

$$A > f(x_n) > A - 1/n.$$

We choose a convergent subsequence from $\{x_n\}:\{x_{n_k}\} \to x$. By continuity, $f(x) = A$. The proof for inf $f(x)$ is entirely analogous.

Theorems 2 and 3 allow generalizations to an even more extensive class of functions (the so-called *semicontinuous functions*).

A function $f(x)$ is said to be *lower (upper) semicontinuous* at the point x_0 if for arbitrary $\epsilon > 0$ there exists a δ-neighborhood of x_0 in which $f(x) > f(x_0) - \epsilon$, $(f(x) < f(x_0) + \epsilon)$.

For example, the function "integral part of x", $f(x) = E(x)$, is upper semicontinuous. If we increase (decrease) the value of $f(x_0)$ of a continuous function at a single point x_0, we obtain a function which is upper (lower) semicontinuous. If $f(x)$ is upper semicontinuous, then $-f(x)$ is lower semicontinuous. These two remarks at once permit us to construct a large number of examples of semicontinuous functions.

We shall also consider functions which assume the values $\pm \infty$. If $f(x_0) = -\infty$, then $f(x)$ will be assumed to be lower semicontinuous at x_0 and upper semicontinuous at x_0 if for arbitrary $h > 0$ there is a neighborhood of the point x_0 in which $f(x) < -h$.

If $f(x_0) = +\infty$, then $f(x)$ will be assumed to be upper semicontinuous at x_0 and lower semicontinuous at x_0 if for arbitrary $h > 0$ there is a neighborhood of the point x_0 in which $f(x) > h$.

The *upper limit* $\bar{f}(x_0)$ of the function $f(x)$ at the point x_0 is the $\lim_{\epsilon \to 0}$ {sup $[f(x); x \in S(x_0, \epsilon)]$}. The *lower limit* $\underline{f}(x_0)$ is the $\lim_{\epsilon \to 0}$ {inf $[f(x); x \in S(x_0, \epsilon)]$}. The difference $\omega f(x_0) = \bar{f}(x_0) - \underline{f}(x_0)$ is the *oscillation* of the function $f(x)$ at the point x_0. It is easy to see that a necessary and sufficient condition that the function $f(x)$ be continuous at the point x_0 is that $\omega f(x_0) = 0$, i.e. that $\bar{f}(x_0) = \underline{f}(x_0)$.

For arbitrary $f(x)$ the function $\bar{f}(x)$ is upper semicontinuous and the function $\underline{f}(x)$ is lower semicontinuous. This follows easily from the definition of the upper and lower limits.

We now consider several important examples of semicontinuous functionals.

Let $f(x)$ be a real function of a real variable. For arbitrary real a and b such that $f(x)$ is defined on the closed interval $[a, b]$ we define the *total variation* of the function $f(x)$ on $[a, b]$ to be the functional

$$V_a^b(f) = \text{sup} \sum_{i=1}^n |f(x_i) - f(x_{i-1})|$$

where $a = x_0 < x_1 < x_2 < \cdots < x_n = b$ and the least upper bound is taken over all possible subdivisions of the closed interval $[a, b]$.

For a monotone function $V_a^b(f) = |f(b) - f(a)|$. For a piecewise monotone function $V_a^b(f)$ is the sum of the absolute values of the increments on the segments of monotonicity. For such functions

$$\sup \sum_{i=1}^{n} | f(x_i) - f(x_{i-1}) |$$

is attained for some subdivision.

We shall prove that the *functional* $V_a^b(f)$ *is lower semicontinuous in the space* M *of all bounded functions of a real variable with metric* $\rho(f, g) = \sup |f(x) - g(x)|$ (it is clear that C is a subspace of the space M), i.e. that for arbitrary f and $\epsilon > 0$ there exists a δ such that $V_a^b(g) > V_a^b(f) - \epsilon$ for $\rho(f, g) < \delta$.

We choose a subdivision of the closed interval $[a, b]$ such that

$$\sum_{i=1}^{n} | f(x_i) - f(x_{i-1}) | > V_a^b(f) - \epsilon/2.$$

Let $\delta = \epsilon/4n$. Then if $\rho(g, f) < \delta$, we have

$$\sum_{i=1}^{n} | f(x_i) - f(x_{i-1}) | - \sum_{i=1}^{n} | g(x_i) - g(x_{i-1}) | < \epsilon/2$$

and consequently

$$V_a^b(g) \geq \sum_{i=1}^{n} | g(x_i) - g(x_{i-1}) | > V_a^b(f) - \epsilon.$$

In the case $V_a^b(f) = \infty$ the theorem remains valid since then for arbitrary H there exists a subdivision of the closed interval $[a, b]$ such that

$$\sum_{i=1}^{n} | f(x_i) - f(x_{i-1}) > H$$

and δ can be chosen such that

$$\sum_{i=1}^{n} | f(x_i) - f(x_{i-1}) | - \sum_{i=1}^{n} | g(x_i) - g(x_{i-1}) | < \epsilon.$$

Then $V_a^b(g) > H - \epsilon$, i.e. $V_a^b(g) \geq H$, so that $V_a^b(g) = \infty$.

The functional $V_a^b(f)$ is not continuous as is easily seen from the following example. Let $f(x) \equiv 0$, $g_n(x) = (1/n) \sin nx$. Then $\rho(g_n, f) = 1/n$, but $V_0^\pi(g_n) = 2$ and $V_0^\pi(f) = 0$.

Functions for which $V_a^b(f) < \infty$ are said to be *functions of bounded* (or finite) *variation*. The reader can find more information about the properties of such functions in the books by Aleksandrov and Kolmogorov: *Introduction to the Theory of Functions of a Real Variable*, Chapter 7, §7; Natanson: *Theory of Functions of a Real Variable*, Chapter 8; and Jeffery: *The Theory of Functions of a Real Variable*, Chapter 5.

We shall define the length of the curve $y = f(x)$ $(a \leq x \leq b)$ as the functional

$$L_a^b(f) = \sup \sum_{i=1}^{n} \{(x_i - x_{i-1})^2 + [f(x_i) - f(x_{i-1})]^2\}^{\frac{1}{2}},$$

where the least upper bound is taken over all possible subdivisions of the closed interval $[a, b]$. This functional is defined on the entire space M. For continuous functions it coincides with the value of the limit

$$\lim \sum_{i=1}^{n} \{(x_i - x_{i-1})^2 + [f(x_i) - f(x_{i-1})]^2\}^{\frac{1}{2}} \text{ as } \max_i | x_i - x_{i-1} | \to 0.$$

Finally, for functions with continuous derivative it can be written in the form

$$\int_a^b [1 + f'^2(x)]^{\frac{1}{2}} \, dx.$$

The functional $L_a^b(f)$ is lower semicontinuous in M. This is proved exactly as in the case of the functional $V_a^b(f)$.

Theorems 2 and 3 established above generalize to semicontinuous functions.

THEOREM 2a. *A finite function which is lower (upper) semicontinuous on a compactum K is bounded below (above) on K.*

In fact, let f be finite and lower semicontinuous and let $\inf f(x) = -\infty$. Then there exists a sequence $\{x_n\}$ such that $f(x_n) < -n$. We choose a subsequence $\{x_{n_k}\} \to x_0$. Then, by virtue of the lower semicontinuity of f, $f(x_0) = -\infty$, which contradicts the assumption that $f(x)$ is finite.

In the case of an upper semicontinuous function the theorem is proved analogously.

THEOREM 3a. *A finite lower (upper) semicontinuous function defined on a compactum K attains its greatest lower (least upper) bound on K.*

Assume the function f is lower semicontinuous. Then by Theorem 2 it has a finite greatest lower bound and there exists a sequence $\{x_n\}$ such that $f(x_n) \leq \inf f(x) + 1/n$. We choose a subsequence $\{x_{n_k}\} \to x_0$. Then $f(x_0) = \inf f(x)$ since the supposition that $f(x_0) > \inf f(x)$ contradicts the lower semicontinuity of f.

The theorem is proved analogously for the case of an upper semicontinuous function.

Let K be a compact metric space and let C_K be the space of continuous real functions defined on K with distance function $\rho(f, g) = \sup |f - g|$. Then the following theorem is valid.

THEOREM 4. *A necessary and sufficient condition that the set $D \subseteq C_K$ be compact is that the functions belonging to D be uniformly bounded and equicontinuous* (Arzelà's theorem for continuous functions defined on an arbitrary compactum).

The sufficiency follows from the general Theorem 7, §18. The necessity is proved exactly as in the proof of Arzelà's theorem (see §17).

§20. Continuous curves in metric spaces

Let $P = f(t)$ be a given continuous mapping of the closed interval $a \leq t \leq b$ into a metric space R. When t runs through the segment from a to b, the corresponding image point P runs through some "continuous curve" in the space R. We propose to give rigorous definitions connected with the above ideas which were stated rather crudely just now. We shall

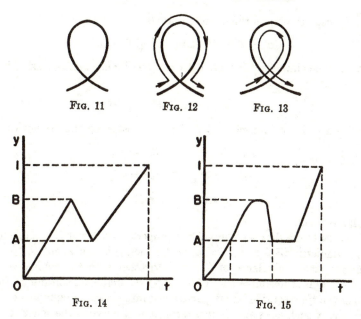

FIG. 11   FIG. 12   FIG. 13

FIG. 14   FIG. 15

consider the order in which the point traverses the curve an essential property of the curve itself. The set shown in Fig. 11, traversed in the directions indicated in Figs. 12 and 13, will be considered as distinct curves. As another example let us consider the real function defined on the closed interval [0, 1] which is shown in Fig. 14. It defines a "curve" situated on the segment [0, 1] of the y-axis, distinct from this segment, traversed once from the point 0 to the point 1, since the segment $[A, B]$ is traversed three times (twice upward and once downward).

However, for the same order of traversing the points of the space we shall consider the choice of the "parameter" t unessential. For example, the functions given in Figs. 14 and 15 define the same "curve" over the y-axis although the values of the parameter t corresponding to an arbitrary point of the curve can be distinct in Figs. 14 and 15. For example, in Fig. 14 to the point A there correspond two isolated points on the t-axis, whereas in Fig. 15 to the point A there correspond on the t-axis one isolated point and the segment lying to the right (when t traverses this segment the point A on the curve remains fixed). [Allowing such intervals of constancy of the point $P = f(t)$ is further convenient in the proof of the compactness of systems of curves.]

We pass over to formal definitions. Two continuous functions

$$P = f_1(t') \quad \text{and} \quad P = f_2(t'')$$

defined, respectively, on the closed intervals

$$a' \leq t' \leq b' \quad \text{and} \quad a'' \leq t'' \leq b''$$

are said to be equivalent if there exist two continuous nondecreasing functions

$$t' = \varphi_1(t) \quad \text{and} \quad t'' = \varphi_2(t)$$

defined on a closed interval $a \leq t \leq b$ and possessing the properties

$$\varphi_1(a) = a', \qquad \varphi_1(b) = b',$$
$$\varphi_2(a) = a'', \qquad \varphi_2(b) = b'',$$
$$f_1[\varphi_1(t)] = f_2[\varphi_2(t)]$$

for all $t \in [a, b]$.

It is easy to see that the equivalence property is reflexive (f is equivalent to f), symmetric (if f_1 is equivalent to f_2, then f_2 is equivalent to f_1), and transitive (the equivalence of f_1 and f_2 together with the equivalence of f_2 and f_3 implies the equivalence of f_1 and f_3). Therefore all continuous functions of the type considered are partitioned into classes of equivalent functions. Every such class also defines a continuous curve in the space R.

It is easy to see that for an arbitrary function $P = f_1(t')$ defined on a closed interval $[a', b']$ we can find a function which is equivalent to it and which is defined on the closed interval $[a'', b''] = [0, 1]$. To this end, it is sufficient to set

$$t' = \varphi_1(t) = (b' - a')t + a', \qquad t'' = \varphi_2(t) = t.$$

(We always assume that $a < b$. However we do not exclude "curves" consisting of a single solitary point which is obtained when the function $f(t)$ is constant on $[a, b]$. This assumption is also convenient in the sequel.) Thus, we can assume that all curves are given parametrically by means of functions defined on the closed interval $[0, 1]$.

Therefore it is expedient to consider the space C_{IR} of continuous mappings of the closed interval $I = [0, 1]$ into the space R with the metric $\rho(f, g) = \sup_t \rho[f(t), g(t)]$.

We shall assume that the sequence of curves $L_1, L_2, \ldots, L_n, \ldots$ converges to the curve L if the curves L_n can be represented parametrically in the form

$$P = f_n(t), \qquad\qquad\qquad 0 \leq t \leq 1,$$

and the curve L in the form

$$P = f(t), \qquad\qquad\qquad 0 \leq t \leq 1,$$

so that $\rho(f, f_n) \to 0$ as $n \to \infty$.

We obtain Theorem 1 if we apply the generalized Arzelà theorem (Theorem 7, §18) to the space C_{IR}.

THEOREM 1. *If the sequence of curves L_1, L_2, $\cdots$, L_n, $\cdots$ lying in the compactum K can be represented parametrically by means of equicontinuous functions defined on the closed interval [0, 1], then this sequence contains a convergent subsequence.*

We shall now define the length of a curve given parametrically by means of the function $P = f(t)$, $a \leq t \leq b$, as the least upper bound of sums of the form

$$\sum_{i=1}^{n} \rho[f(t_{i-1}), f(t_i)],$$

where the points t_i are subject only to the following conditions:

$$a \leq t_0 \leq t_1 \leq \cdots \leq t_i \leq \cdots \leq t_n = b.$$

It is easy to see that the length of a curve does not depend on the choice of its parametric representation. If we limit ourselves to parametric representations by functions defined on the closed interval [0, 1], then it is easy to prove by considerations similar to those of the preceding section that the length of a curve is a lower semicontinuous functional of f (in the space C_{IR}). In geometric language this result can be expressed in the form of such a theorem on semicontinuity.

THEOREM 2. *If the sequence of curves L_n converges to the curve L, then the length of L is not greater than the greatest lower bound of the lengths of the curves L_n.*

We shall now consider specially *curves of finite length* or *rectifiable curves.* Let the curve be defined parametrically by means of the function $P = f(t)$, $a \leq t \leq b$. The function f, considered only on the closed interval $[a, T]$, where $a \leq T \leq b$, defines an "initial segment" of the curve from the point $P_a = f(a)$ to the point $P_T = f(T)$. Let $s = \varphi(T)$ be its length. It is easily established that

$$P = g(s) = f[\varphi^{-1}(s)]$$

is a new parametric representation of the same curve. In this connection s runs through the closed interval $0 \leq s \leq S$, where S is the length of the entire curve under consideration. This representation satisfies the requirement

$$\rho[g(s_1), g(s_2)] \leq |s_2 - s_1|$$

(the length of the curve is not less than the length of the chord).

Going over to the closed interval [0, 1] we obtain the parametric representation

$$P = F(\tau) = g(s), \qquad \tau = s/S$$

which satisfies the following Lipschitz condition:

$$\rho[F(\tau_1), F(\tau_2)] \leq S \mid \tau_1 - \tau_2 \mid.$$

We thus see that for all curves of length S such that $S \leq M$, where M is a constant, a parametric representation on the closed interval $[0, 1]$ by means of equicontinuous functions is possible. Consequently, Theorem 1 is applicable to such curves.

We shall show the power of the general results obtained above by applying them to the proof of the following important proposition.

THEOREM 3. *If two points A and B in the compactum K can be connected by a continuous curve of finite length, then among all such curves there exists one of minimal length.*

In fact, let Y be the greatest lower bound of the lengths of curves which connect A and B in the compactum K. Let the lengths of the curves L_1, L_2, $\cdots$, L_n, $\cdots$ connecting A with B tend to Y. From the sequence L_n it is possible, by Theorem 1, to select a convergent subsequence. By Theorem 2 the limit curve of this subsequence cannot have length greater than Y.

We note that even when K is a closed smooth (i.e. differentiable a sufficient number of times) surface in three-dimensional Euclidean space, this theorem does not follow directly from the results established in usual differential geometry courses where we restrict ourselves ordinarily to the case of sufficiently proximate points A and B.

All the arguments above would take on great clarity if we formed of the set of all curves of a given metric space R a metric space. This can be done by introducing the distance between two curves L_1 and L_2 by means of the formula

$$\rho(L_1, L_2) = \inf \rho(f_1, f_2),$$

where the greatest lower bound is taken over all possible pairs of parametric representations

$$P = f_1(t), \qquad P = f_2(t) \qquad\qquad (0 \leq t \leq 1)$$

of the curves L_1 and L_2, respectively.

The proof of the fact that this distance satisfies the axioms of a metric space is very straightforward with the exception of one point: there is some difficulty in proving that $\rho(L_1, L_2) = 0$ implies that the curves L_1 and L_2 are identical. This fact is an immediate consequence of the fact that the greatest lower bound in the formula which we used in the definition of the distance $\rho(L_1, L_2)$ is attained for a suitable choice of the parametric representations f_1 and f_2. But the proof of this last assertion is also not very straightforward.

Chapter III
NORMED LINEAR SPACES

§21. Definition and examples of normed linear spaces

DEFINITION 1. A set R of elements $x, y, z, \cdots$ is said to be a *linear space* if the following conditions are satisfied:

I. For any two elements $x, y \in R$ there is uniquely defined a third element $z = x + y$, called their sum, such that

1) $x + y = y + x$,
2) $x + (y + z) = (x + y) + z$,
3) there exists an element 0 having the property that $x + 0 = x$ for all $x \in R$, and
4) for every $x \in R$ there exists an element $-x$ such that $x + (-x) = 0$.

II. For an arbitrary number α and element $x \in R$ there is defined an element αx (the product of the element x and the number α) such that

1) $\alpha(\beta x) = (\alpha\beta)x$, and
2) $1 \cdot x = x$.

III. The operations of addition and multiplication are related in the following way:

1) $(\alpha + \beta)x = \alpha x + \beta x$, and
2) $\alpha(x + y) = \alpha x + \alpha y$.

Depending on the numbers admitted (all complex numbers or only the reals), we distinguish between complex and real linear spaces. Unless otherwise stated we shall consider real linear spaces. In a linear space, besides the operations of addition and multiplication by scalars, usually there is introduced in one way or another the operation of passage to the limit. It is most convenient to do this by introducing a norm into the linear space.

A linear space R is said to be *normed* if to each element $x \in R$ there is made to correspond a nonnegative number $\| x \|$ which is called the norm of x and such that:

1) $\| x \| = 0$ if, and only if, $x = 0$,
2) $\| \alpha x \| = | \alpha | \| x \|$,
3) $\| x + y \| \leq \| x \| + \| y \|$.

It is easy to see that every normed space is also a metric space; it is sufficient to set $\rho(x, y) = \| x - y \|$. The validity of the metric space axioms follows directly from Properties 1–3 of the norm.

A complete normed space is said to be a *Banach space, a space of Banach type*, or, more briefly, a *B-space*.

EXAMPLES OF NORMED SPACES. 1. The real line with the usual arithmetic definitions is the simplest example of a normed space. In this case the norm is simply the absolute value of the real number.

2. Euclidean n-space, i. e. the space consisting of all n-tuples of real numbers: $x = (x_1, x_2, \cdots, x_n)$ in which the norm (i. e. the length) of the vector is defined to be the square root of its scalar square,

$$\| x \| = \left(\sum_{i=1}^{n} x_i^2 \right)^{\frac{1}{2}},$$

is also a normed linear space.

In an n-dimensional linear space the norm of the vector $x = (x_1, x_2, \cdots, x_n)$ can also be defined by means of the formula

$$\| x \| = \left(\sum_{k=1}^{n} | x_k |^p \right)^{1/p}, \qquad (p \geq 1).$$

We also obtain a normed space if we set the norm of the vector $x = (x_1, x_2, \cdots, x_n)$ equal to the max $\{ |x_k| ; 1 \leq k \leq n \}$.

3. The space $C[a, b]$ of continuous functions with the operations of addition and multiplication by a scalar which are usual for functions, in which

$$\| f(t) \| = \max \{ | f(t) | ; a \leq t \leq b \},$$

is a normed linear space.

4. Let $C^2[a, b]$ consist of all functions continuous on $[a, b]$ and let the norm be given by the formula

$$\| f(t) \| = \left(\int_a^b f^2(t) \, dt \right)^{\frac{1}{2}}.$$

All the norm axioms are satisfied.

5. The space l_2 is a normed linear space if we define the sum of two elements $x = (\xi_1, \xi_2, \cdots, \xi_n, \cdots)$ and $y = (\eta_1, \eta_2, \cdots, \eta_n, \cdots)$ in l_2 to be

$$x + y = (\xi_1 + \eta_1, \xi_2 + \eta_2, \cdots, \xi_n + \eta_n, \cdots),$$

and let

$$\alpha x = (\alpha \xi_1, \alpha \xi_2, \cdots, \alpha \xi_n, \cdots),$$

and

$$\| x \| = \left(\sum_{n=1}^{\infty} | \xi_n |^2 \right)^{\frac{1}{2}}.$$

6. The space c consisting of all sequences $x = (x_1, x_2, \cdots, x_n, \cdots)$ of real numbers which satisfy the condition $\lim_{n \to \infty} x_n = 0$.

Addition and multiplication are defined as in Example 5 and the norm is set equal to

$$\| x \| = \max \{ | x_n | ; 1 \leq n \leq \infty \}.$$

7. The space m of bounded sequences with the same definitions of sum, product, and norm as in the preceding example.

In each of these examples the linear space axioms are verified without difficulty. The fact that the norm Axioms 1–3 are fulfilled in Examples 1–5 is proved exactly as was the validity of the metric space axioms in the corresponding examples in §8, Chapter II.

All the spaces enumerated in the examples, except the space $C^2[a, b]$, are Banach spaces.

DEFINITION 2. A *linear manifold* L in a normed linear space R is any set of elements in R satisfying the following condition: if $x, y \in L$, then $\alpha x + \beta y \in L$, where α and β are arbitrary numbers. A *subspace* of the space R is a closed linear manifold in R.

REMARK 1. In Euclidean n-space R^n the concepts of linear manifold and subspace coincide because every linear manifold in R^n is automatically closed. (Prove this!) On the other hand, linear manifolds which are not closed exist in an infinite-dimensional space. For example, in l_2 the set L of points of the form

(1) $$x = (x_1, x_2, \cdots, x_k, 0, 0, \cdots),$$

i.e. of points which have only a finite (but arbitrary) number of nonzero coordinates, forms a linear manifold which is not closed. In fact, a linear combination of points of form (1) is a point of the same form, i.e. L is a linear manifold. But L is not closed since, for instance, the sequence of points

$$(1, 0, 0, 0, \cdots),$$
$$(1, \tfrac{1}{2}, 0, 0, \cdots),$$
$$(1, \tfrac{1}{2}, \tfrac{1}{4}, 0, \cdots),$$
$$\cdots\cdots\cdots\cdots,$$

belonging to L, converges to the point $(1, \tfrac{1}{2}, \tfrac{1}{4}, \cdots, 1/2^n, \cdots)$, which does not belong to L.

REMARK 2. Let $x_1, x_2, \cdots, x_n, \cdots$ be elements of a Banach space R and let M be the totality of elements in R which are of the form $\sum_{i=1}^{n} c_i x_i$ for arbitrary finite n. It is obvious that M is a linear manifold in R. We shall show that $[M]$ is a linear subspace. In view of the fact that $[M]$ is closed it is sufficient to prove that it is a linear manifold.

Let $x \in [M], y \in [M]$. Then in an arbitrary ϵ-neighborhood of x we can find an $x_\epsilon \in M$ and in an arbitrary ϵ-neighborhood of y we can find a $y_\epsilon \in M$. We form the element $\alpha x + \beta y$ and estimate $\| \alpha x + \beta y - \alpha x_\epsilon - \beta y_\epsilon \|$:

$$\| \alpha x + \beta y - \alpha x_\epsilon - \beta y_\epsilon \|$$
$$\leq | \alpha | \| x - x_\epsilon \| + | \beta | \| y - y_\epsilon \| \leq (| \alpha | + | \beta |)\epsilon,$$

from which it is clear that $\alpha x + \beta y \in [M]$.

The subspace $L = [M]$ is said to be the *subspace generated by the elements* $x_1, x_2, \cdots, x_n, \cdots$.

§22. Convex sets in normed linear spaces

Let x and y be two points in the linear space R. Then the *segment* connecting the points x and y is the totality of all points of the form $\alpha x + \beta y$, where $\alpha \geq 0$, $\beta \geq 0$, and $\alpha + \beta = 1$.

DEFINITION. A set M in the linear space R is said to be *convex* if, given two arbitrary points x and y belonging to M, the segment connecting them also belongs to M. A convex set is called a *convex body* if it contains at least one interior point, i.e. if it contains some sphere completely.

EXAMPLES. 1. In three-dimensional Euclidean space, the cube, sphere, tetrahedron, and halfspace are convex bodies; but a triangle, plane, and segment are convex sets although they are not convex bodies.

2. A sphere in a normed linear space is always a convex set (and also a convex body). In fact, consider the unit sphere S: $\| x \| \leq 1$.

If x_0, y_0 are two arbitrary points belonging to this sphere: $\| x_0 \| \leq 1$, $\| y_0 \| \leq 1$, then

$$\| \alpha x_0 + \beta y_0 \| \leq \| \alpha x_0 \| + \| \beta y_0 \| = \alpha \| x_0 \| + \beta \| y_0 \| \leq \alpha + \beta = 1,$$

i.e.

$$\alpha x_0 + \beta y_0 \in S \qquad (\alpha \geq 0, \beta \geq 0, \alpha + \beta = 1).$$

3. Let R be the totality of vectors $x = (\xi_1, \xi_2)$ in the plane. Introduce the following distinct norms in R:

$$\| x \|_2 = (\xi_1^2 + \xi_2^2)^{\frac{1}{2}}; \qquad \| x \|_\infty = \max(\, | \xi_1 |, | \xi_2 | \,);$$

$$\| x \|_1 = | \xi_1 | + | \xi_2 |; \qquad \| x \|_p = (\, | \xi_1 |^p + | \xi_2 |^p \,)^{1/p} \quad (p > 1).$$

Let us see what the unit sphere will be for each of these norms (see Fig. 16).

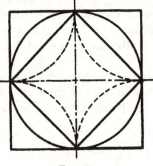

FIG. 16

In the case $\| x \|_2$ it is a circle of radius 1, in the case $\| x \|_\infty$ it is a square with vertices $(\pm 1, \pm 1)$, in the case $\| x \|_1$ it is a square with vertices $(0, 1)$, $(1, 0)$, $(-1, 0)$, $(0, -1)$. If we consider the unit sphere corresponding to the norm $\| x \|_p$, and let p increase from 1 to ∞, then this "sphere" deforms in a continuous manner from the square corresponding to $\| x \|_1$ to the square corresponding to $\| x \|_\infty$. Had we set

$$(1) \qquad \| x \|_p = (| \xi_1 |^p + | \xi_2 |^p)^{1/p}$$

for $p < 1$, then the set $\| x \|_p \leq 1$ would not have been convex (for example, for $p = 2/3$ it would be the interior of an astroid). This is another expression of the fact that for $p < 1$ the "norm" (1) does not satisfy Condition 3 in the definition of a norm.

4. Let us consider a somewhat more complicated example. Let Φ be the set of points $x = (\xi_1, \xi_2, \cdots, \xi_n, \cdots)$ in l_2 which satisfy the condition

$$\sum_{n=1}^\infty \xi_n^2 n^2 \leq 1.$$

This is a convex set in l_2 which is not a convex body. In fact, if $x, y \in \Phi$ and $z = \alpha x + \beta y$, where $\alpha, \beta \geq 0$ and $\alpha + \beta = 1$, then by virtue of the Schwarz inequality (Chapter II),

$$\sum_{n=1}^\infty n^2(\alpha\xi_n + \beta\eta_n)^2 = \alpha^2 \sum_{n=1}^\infty n^2\xi_n^2 + 2\alpha\beta \sum_{n=1}^\infty n^2\xi_n\eta_n + \beta^2 \sum_{n=1}^\infty n^2\eta_n^2$$
$$\leq \alpha^2 \sum_{n=1}^\infty n^2\xi_n^2 + 2\alpha\beta (\sum_{n=1}^\infty n^2\xi_n^2)^{\frac{1}{2}}(\sum_{n=1}^\infty n^2\eta_n^2)^{\frac{1}{2}} + \beta^2 \sum_{n=1}^\infty n^2\eta_n^2$$
$$= [\alpha(\sum n^2\xi_n^2)^{\frac{1}{2}} + \beta(\sum n^2\eta_n^2)^{\frac{1}{2}}]^2 \leq (\alpha + \beta)^2 = 1.$$

We shall show that Φ contains no sphere. Φ is symmetric with respect to the origin of coordinates; hence, if Φ contained some sphere S', it would also contain the sphere S'' which is symmetric to S' with respect to the origin. Then Φ, being convex, would contain all segments connecting points of the spheres S' and S'', and consequently it would also contain a sphere S of the same radius as that of S', with the center of S at the origin. But if Φ contained some sphere of radius r with center at the origin, then on every ray emanating from zero there would lie a segment belonging entirely to Φ. However, on the ray defined by the vector $(1, 1/2, 1/3, \cdots, 1/n, \cdots)$ there obviously is no point except zero which belongs to Φ.

EXERCISES. 1. Prove that the set Φ is compact. Prove that no compact convex set in l_2 can be a convex body.

2. Prove that Φ is not contained in any subspace distinct from all of l_2.

3. Prove that the fundamental parallelopiped in l_2 (see Example 3, §16) is a convex set but not a convex body.

We shall now establish the following simple properties of convex sets.

THEOREM 1. *The closure of a convex set is a convex set.*

Proof. Let M be a convex set, $[M]$ its closure and let x, y be two arbitrary points in $[M]$. Further, let ϵ be an arbitrary positive number. Points a, b can be found in M such that $\rho(a, x) < \epsilon$ and $\rho(b, y) < \epsilon$. Then $\rho(\alpha x + \beta y, \alpha a + \beta b) < \epsilon$ for arbitrary nonnegative α and β such that $\alpha + \beta = 1$, and the point $\alpha a + \beta b$ belongs to M since M is convex. Since $\epsilon > 0$ is arbitrary, it follows that $\alpha x + \beta y \in [M]$, i. e. $[M]$ is convex also.

THEOREM 2. *The intersection of an arbitrary number of convex sets is a convex set.*

Proof. Let $M = \bigcap_\alpha M_\alpha$, where all M_α are convex sets. Further, let x and y be two arbitrary points in M. These points x and y belong to all M_α. Then the segment connecting the points x and y belongs to each M_α and consequently it also belongs to M. Thus, M is in fact convex.

Since the intersection of closed sets is always closed, it follows that *the intersection of an arbitrary number of closed convex sets is a closed convex set.*

Let A be an arbitrary subset of a normed linear space. We define the *convex closure* of the set A to be the smallest closed convex set containing A.

The convex closure of any set can obviously be obtained as the intersection of all closed convex sets which contain the given set.

Consider the following important example of convex closure. Let x_1, x_2, $\cdots$, x_{n+1} be points in a normed linear space. We shall say that these $n + 1$ points are *in general position* if no three of them lie on one straight line, no four of them lie in one plane, and so forth; in general, no $k + 1$ of these points lie in a subspace of dimension less than k. The convex closure of the points x_1, x_2, $\cdots$, x_{n+1} which are in general position is called an *n-dimensional simplex* and the points x_1, x_2, $\cdots$, x_{n+1} themselves are called the *vertices* of the simplex. A zero-dimensional simplex consists of a single point. One-, two-, and three-dimensional simplexes are, respectively, a segment, triangle, tetrahedron.

If the points x_1, x_2, $\cdots$, x_{n+1} are in general position, then any $k + 1$ of them ($k < n$) also are in general position and consequently they generate a k-dimensional simplex, called a *k-dimensional face* of the given n-dimensional simplex. For example, the tetrahedron with the vertices e_1, e_2, e_3, e_4 has four two-dimensional faces defined respectively by the triples of vertices (e_2, e_3, e_4), (e_1, e_3, e_4), (e_1, e_2, e_4), (e_1, e_2, e_3); six one-dimensional faces; and four zero-dimensional faces.

THEOREM 3. *A simplex with the vertices x_1, x_2, $\cdots$, x_{n+1} is the totality of all points which can be represented in the form*

$$(2) \qquad x = \sum_{k=1}^{n+1} \alpha_k x_k ; \qquad \alpha_k \geq 0, \qquad \sum_{k=1}^{n+1} \alpha_k = 1.$$

Proof. In fact, it is easy to verify that the totality of points of the form (2) represents a closed convex set which contains the points x_1, x_2, $\cdots$, x_{n+1}. On the other hand, every convex set which contains the points x_1, x_2,

$\cdots$, x_{n+1} must also contain points of the form (2), and consequently these points form the smallest closed convex set containing the points x_1, x_2, $\cdots$, x_{n+1} .

§23. Linear functionals

DEFINITION 1. A numerical function $f(x)$ defined on a normed linear space R will be called a *functional*. A functional $f(x)$ is said to be *linear* if

$$f(\alpha x + \beta y) = \alpha f(x) + \beta f(y),$$

where x, $y \in R$ and α, β are arbitrary numbers.

A functional $f(x)$ is said to be *continuous* if for arbitrary $\epsilon > 0$ a $\delta > 0$ can be found such that the inequality

$$| f(x_1) - f(x_2) | < \epsilon$$

holds whenever

$$\| x_1 - x_2 \| < \delta.$$

In the sequel we shall consider only continuous functionals (in particular continuous linear functionals) and for brevity we shall omit the word "continuous".

We shall establish some properties of linear functionals which follow almost directly from the definition.

THEOREM 1. *If the linear functional $f(x)$ is continuous at some point $x_0 \in R$, then it is continuous everywhere in R.*

Proof. In fact, let the linear functional $f(x)$ be continuous at the point $x = x_0$. This is equivalent to the fact that $f(x_n) \to f(x_0)$ when $x_n \to x_0$.

Further, let $y_n \to y$. Then

$$f(y_n) = f(y_n - y + x_0 + y - x_0) = f(y_n - y + x_0) + f(y) - f(x_0).$$

But $y_n - y + x_0 \to x_0$. Consequently, by assumption, $f(y_n - y + x_0) \to f(x_0)$. Thus,

$$f(y_n) \to f(x_0) + f(y) - f(x_0) = f(y).$$

A functional $f(x)$ is said to be *bounded* if there exists a constant N such that

(1) $$| f(x) | < N \| x \|$$

for all $x \in R$.

THEOREM 2. *For linear functionals the conditions of continuity and boundedness are equivalent.*

Proof. We assume that the linear functional $f(x)$ is not bounded. Then for arbitrary natural number n we can find an element $x_n \in R$ such that $|f(x_n)| > n \| x_n \|$. We shall set $y_n = x_n/(n \| x_n \|)$. Then $\| y_n \| = 1/n$, i.e. $y_n \to 0$. But at the same time

$$|f(y_n)| = |f(x_n/n \| x_n \|)| = (1/n \| x_n \|) |f(x_n)| > 1.$$

Consequently, the functional $f(x)$ is not continuous at the point $x = 0$.

Now let N be a number which satisfies Condition (1). Then for an arbitrary sequence $x_n \to 0$ we have:

$$|f(x_n)| \leq N \| x_n \| \to 0,$$

i.e. $f(x)$ is continuous at the point $x = 0$ and consequently at all the remaining points also. This completes the proof of the theorem.

DEFINITION 2. The quantity

$$\| f \| = \sup \{ |f(x)|/\| x \|; x \neq 0 \}$$

is called the *norm* of the linear functional $f(x)$.

EXAMPLES OF LINEAR FUNCTIONALS ON VARIOUS SPACES. 1. Let R^n be Euclidean n-space and let a be a fixed nonzero vector in R^n. For arbitrary $x \in R^n$ we set $f(x) = (x, a)$, where (x, a) is the scalar product of the vectors x and a. It is clear that $f(x)$ is a linear functional. In fact,

$$f(\alpha x + \beta y) = (\alpha x + \beta y, a) = \alpha(x, a) + \beta(y, a) = \alpha f(x) + \beta f(y).$$

Further, by virtue of Schwarz's inequality

(2) $$|f(x)| = |(x, a)| \leq \| x \| \| a \|.$$

Consequently, the functional $f(x)$ is bounded and is therefore continuous. From (2) we find

$$|f(x)|/\| x \| \leq \| a \|.$$

Since the right member of this inequality does not depend on x, we have

$$\sup |f(x)|/\| x \| \leq \| a \|,$$

i.e. $\| f \| \leq \| a \|$. But, setting $x = a$ we obtain:

$$|f(a)| = (a, a) = \| a \|^2, \quad \text{i.e.} \quad (|f(a)|/\| a \|) = \| a \|.$$

Therefore $\| f \| = \| a \|$.

If a is zero, then f is the zero linear functional. Hence $\| f \| = \| a \|$ in this case also.

2. The integral

$$I = \int_a^b x(t) \, dt$$

($x(t)$ is a continuous function on $[a, b]$) represents a linear functional on the space $C[a, b]$. Its norm equals $b - a$. In fact,

$$| I | = \left| \int_a^b x(t)\ dt \right| \leq \max | x(t) | (b - a),$$

where equality holds when $x =$ constant.

3. Now let us consider a more general example. Let $y_0(t)$ be a fixed continuous function on $[a, b]$. We set, for arbitrary function $x(t) \in C[a, b]$,

$$f(x) = \int_a^b x(t)y_0(t)\ dt.$$

This expression represents a linear functional on $C[a, b]$ because

$$f(\alpha x + \beta y) = \int_a^b (\alpha x(t) + \beta y(t))y_0(t)\ dt$$

$$= \alpha \int_a^b x(t)y_0(t)\ dt + \beta \int_a^b y(t)y_0(t)\ dt = \alpha f(x) + \beta f(y).$$

This functional is bounded. In fact,

$$| f(x) | = \left| \int_a^b x(t)y_0(t)\ dt \right| \leq \| x \| \int_a^b | y_0(t) | \ dt.$$

Thus, the functional $f(x)$ is linear and bounded and consequently it is continuous also. It is possible to show that its norm is exactly equal to $\int_a^b | y_0(t) | \ dt$.

4. We now consider on the same space $C[a, b]$ a linear functional of another type, namely, we set

$$\delta_{t_0}x(t) = x(t_0),$$

i.e. the value of the functional δ_{t_0} for the function $x(t)$ is equal to the value of this function at the fixed point t_0. This functional is frequently encountered, for example, in quantum mechanics where it is usually written in the form

$$\delta_{t_0}x(t) = \int_a^b x(t)\delta(t - t_0)\ dt,$$

where $\delta(t)$ is the "function" equal to zero everywhere except at the point $t = 0$ and such that its integral equals unity (the Dirac δ-function). The δ-function can be thought of as the limit, in some sense, of a sequence of functions $\varphi_n(t)$ each of which assumes the value zero outside some ϵ_n-neighborhood ($\epsilon_n \to 0$ as $n \to \infty$) of the point $t = 0$ and such that the integral of the limiting function equals 1.

5. In the space l_2 we can define a linear functional as in R^n by choosing in l_2 some fixed element $a = (a_1, a_2, \cdots, a_n, \cdots)$ and setting

$$(3) \qquad\qquad f(x) = \sum_{n=1}^{\infty} x_n a_n .$$

Series (3) converges for arbitrary $x \in l_2$ and

$$(4) \qquad | \sum_{n=1}^{\infty} x_n a_n | \leq (\sum_{n=1}^{\infty} x_n^2)^{\frac{1}{2}} (\sum_{n=1}^{\infty} a_n^2)^{\frac{1}{2}} = \| x \| \, \| a \|.$$

Inequality (4) transforms into the identity $\sum_{n=1}^{\infty} a_n^2 \equiv \sum_{n=1}^{\infty} a_n^2$ for $x = a$ and consequently $\| f \| = \| a \|$.

Geometric interpretation of a linear functional. Let $f(x)$ be a linear functional on the space R. We shall assume $f(x)$ is not identically zero. The set L_f of those elements x in R which satisfy the condition $f(x) = 0$ form a subspace. In fact, if $x, y \in L_f$, then

$$f(\alpha x + \beta y) = \alpha f(x) + \beta f(y) = 0,$$

i.e. $\alpha x + \beta y \in L_f$. Further, if $x_n \to x$ and $x_n \in L_f$, then by virtue of the continuity of the functional f,

$$f(x) = \lim_{n \to \infty} f(x_n) = 0.$$

DEFINITION 3. We say that the subspace L of the Banach space R has *index* (or *deficiency*) s if: 1) R contains s linearly independent elements $x_1, x_2, \cdots, x_s$ which do not belong to L with the property that every element $x \in R$ can be represented in the form

$$x = a_1 x_1 + a_2 x_2 + \cdots + a_s x_s + y, \qquad y \in L;$$

and 2) it is impossible to find a smaller number of elements x_i which possess the indicated properties.

In the case of a finite-dimensional space R the index plus the dimension of the subspace L is equal to the dimension of the whole space.

THEOREM 3. *Let $f(x) \neq 0$ be a given functional. The subspace L_f has index equal to unity, i.e. an arbitrary element $y \in R$ can be represented in the form*

$$(5) \qquad\qquad y = \lambda x_0 + x,$$

where $x \in L_f$, $x_0 \notin L_f$.

Proof. Since $x_0 \notin L_f$, we have $f(x_0) \neq 0$. If we set $\lambda = f(y)/f(x_0)$ and $x = y - \{f(y)/f(x_0)\}x_0$, then $y = \lambda x_0 + x$, where

$$f(x) = f(y) - (f(y)/f(x_0))f(x_0) = 0.$$

If the element x_0 is fixed, then the element y can be represented in the form (5) uniquely. This is easily proved by assuming the contrary. In fact, let

$$y = \lambda x_0 + x,$$
$$y = \lambda' x_0 + x';$$

then

$$(\lambda - \lambda')x_0 = (x' - x).$$

If $\lambda - \lambda' = 0$, then obviously, $x - x' = 0$. But if $\lambda - \lambda' \neq 0$, then $x_0 = (x' - x)/(\lambda - \lambda') \in L_f$, which contradicts the condition that $x_0 \notin L_f$.

Conversely, given a subspace L of R of index 1, L defines a continuous linear functional f which vanishes precisely on L. Indeed, let $x_0 \notin L$. Then for any $x \in R$, $x = y + \lambda x_0$, with $y \in L$, $x_0 \notin L$. Let $f(x) = \lambda$. It is easily seen that f satisfies the above requirements. If f, g are two such linear functionals defined by L, then $f(x) = \alpha g(x)$ for all $x \in R$, α a scalar. This follows because the index of L in R is 1.

We shall now consider the totality M_f of elements in R which satisfy the condition $f(x) = 1$. M_f can be represented in the form $M_f = L_f + x_0$, where x_0 is a fixed element such that $f(x_0) = 1$ and L_f is the totality of elements which satisfy the condition $f(x) = 0$. In analogy with the finite-dimensional case it is natural to call M_f a *hyperplane* in the space R. It is easy to verify that the hyperplanes $f(x) = 1$ and $\varphi(x) = 1$ coincide if, and only if, the functionals f and φ coincide. Thus, it is possible to establish a one-to-one correspondence between all functionals defined on R and all hyperplanes in R which do not pass through the origin of coordinates.

We shall now find the distance from the hyperplane $f(x) = 1$ to the origin. It is equal to

$$d = \inf \{\| x \|; f(x) = 1\}.$$

For all x such that $f(x) = 1$ we have

$$1 \leq \| f \| \| x \|, \quad \text{i.e. } \| x \| \geq 1/\| f \|;$$

therefore $d \geq 1/\| f \|$. Further, since for arbitrary $\epsilon > 0$ an element x satisfying the condition $f(x) = 1$ can be found such that

$$1 > (\| f \| - \epsilon) \| x \|,$$

it follows that

$$d = \inf \{\| x \| < 1/(\| f \| - \epsilon); f(x) = 1\}.$$

Consequently,

$$d = 1/\| f \|,$$

i.e. *the norm of the linear functional $f(x)$ equals the reciprocal of the magnitude of the distance of the hyperplane $f(x) = 1$ from the origin of coordinates.*

§24. The conjugate space

It is possible to define the operations of addition and multiplication by a scalar for linear functionals. Let f_1 and f_2 be two linear functionals on a normed linear space R. Their sum is a linear functional $f = f_1 + f_2$ such that $f(x) = f_1(x) + f_2(x)$ for arbitrary $x \in R$.

The product of a linear functional f_1 by a number α is a functional $f = \alpha f_1$ such that

$$f(x) = \alpha f_1(x)$$

for arbitrary $x \in R$.

It is easy to verify that the operations of addition and multiplication by a scalar of functionals so defined satisfy all the axioms of a linear space. Moreover, the definition we gave above of the norm of a linear functional satisfies all the requirements found in the definition of a normed linear space. In fact,

1) $\| f \| > 0$ for arbitrary $f \not\equiv 0$,

2) $\| \alpha f \| = | \alpha | \| f \|$,

3) $\| f_1 + f_2 \| = \sup \{| f_1(x) + f_2(x) |/\| x \|\}$

$\leq \sup \{(| f_1(x) | + | f_2(x) |)/\| x \|\}$

$\leq \sup \{| f_1(x) |/\| x \|\} + \sup \{| f_2(x) |/\| x \|\}$

$= \| f_1 \| + \| f_2 \|.$

Thus, the totality of all linear functionals on a normed space R itself represents a normed linear space; it is called the conjugate space of R and is denoted by $\bar{R}$.

THEOREM 1. *The conjugate space is always complete.*

Proof. Let $\{f_n\}$ be a fundamental sequence of linear functionals. By the definition of a fundamental sequence, for every $\epsilon > 0$ there exists an N such that $\| f_n - f_m \| < \epsilon$ for all $n, m > N$. Then for arbitrary $x \in R$,

$$| f_n(x) - f_m(x) | \leq \| f_n - f_m \| \| x \| < \epsilon \| x \|,$$

i.e. for arbitrary $x \in R$ the numerical sequence $f_n(x)$ converges.

If we set

$$f(x) = \lim_{n\to\infty} f_n(x),$$

then $f(x)$ represents a linear functional. In fact,

1) $f(\alpha x + \beta y) = \lim_{n\to\infty} f_n(\alpha x + \beta y)$

$= \lim_{n\to\infty} [\alpha f_n(x) + \beta f_n(y)] = \alpha f(x) + \beta f(y).$

2) Choose N so that $\| f_n - f_{n+p} \| < 1$ for all $n > N$. Then

$$\| f_{n+p} \| < \| f_n \| + 1$$

for all p. Consequently, $| f_{n+p}(x) | \leq (\| f_n \| + 1) \| x \|.$

Passing to the limit as $p \to \infty$, we obtain

$$\lim_{p \to \infty} |f_{n+p}(x)| = |f(x)| \leq (\|f_n\| + 1)\|x\|,$$

i.e. the functional $f(x)$ is bounded. We shall now prove that the functional f is the limit of the sequence $f_1, f_2, \cdots, f_n, \cdots$. By the definition of the norm, for every $\epsilon > 0$ there exists an element x_ϵ such that

$$\|f_n - f\| \leq \{(|f_n(x_\epsilon) - f(x_\epsilon)|)/\|x_\epsilon\|\} + \epsilon/2$$
$$= |f_n(x_\epsilon/\|x_\epsilon\|) - f(x_\epsilon/\|x_\epsilon\|)| + \epsilon/2;$$

since

$$f(x_\epsilon/\|x_\epsilon\|) = \lim_{n \to \infty} f_n(x_\epsilon/\|x_\epsilon\|),$$

it is possible to find an $n_0(\epsilon)$ such that for $n > n_0$

$$|f_n(x_\epsilon/\|x_\epsilon\|) - f(x_\epsilon/\|x_\epsilon\|)| < \epsilon/2,$$

so that for $n > n_0$ the inequality

$$\|f_n - f\| < \epsilon$$

is fulfilled.

This completes the proof of the theorem.

Let us emphasize once more that this theorem is valid independently of whether the initial space R is complete or not.

EXAMPLES. 1. Let the space E be finite-dimensional with basis $e_1, e_2, \cdots,$ e_n. Then the functional $f(x)$ is expressible in the form

$$(1) \qquad f(x) = \sum_{i=1}^{n} f_i x_i,$$

where $x = \sum_{i=1}^{n} x_i e_i$ and $f_i = f(e_i)$.

Thus, the functional is defined by the n numbers $f_1, \cdots, f_n$ which are the values of f on the basis vectors. The space which is the conjugate of the finite-dimensional space is also finite-dimensional and has the same dimension.

The explicit form assumed by the norm in the conjugate space depends on the choice of norm in E.

a) Let $\|x\| = (\sum x_i^2)^{\frac{1}{2}}$. We have already shown that then

$$\|f\| = (\sum f_i^2)^{\frac{1}{2}},$$

i.e. the conjugate of an Euclidean space is itself Euclidean.

b) Let $\|x\| = \sup_i |x_i|$. Then

$$|f(x)| = |\sum f_i x_i| \leq (\sum |f_i|) \sup_i |x_i| = (\sum |f_i|)\|x\|.$$

From this it follows that

$$\|f\| \leq \sum |f_i|.$$

The norm $\| f \|$ cannot be less than $\sum |f_i|$ since if we set

$$x_i = \begin{cases} +1 \text{ if } f_i > 0, \\ -1 \text{ if } f_i < 0, \\ 0 \text{ if } f_i = 0, \end{cases}$$

then the following equality is valid:

$$|f(x)| = \sum |f_i| = \left(\sum |f_i|\right) \| x \|.$$

c) If $\| x \| = \left(\sum |x_i|^p\right)^{1/p}$, $p > 1$, then $\| f \| = \left(\sum |f_i|^q\right)^{1/q}$, where $1/p + 1/q = 1$. This follows from the Hölder inequality

$$\left| \sum f_i x_i \right| \le \left(\sum |x_i|^p\right)^{1/p} \left(\sum |f_i|^q\right)^{1/q}$$

and from the fact that the equality sign is attained [for $f_i = (\operatorname{sgn} x_i)(x_i)^{p-1}$].

2. Let us consider the space c consisting of sequences $x = (x_1, x_2, \cdots, x_n, \cdots)$ which are such that $x_n \to 0$ as $n \to \infty$, where $\| x \| = \sup_n x_n$.

If a functional in the space c is expressible by means of the formula

$$(2) \qquad f(x) = \sum_{i=1}^\infty f_i x_i, \qquad \sum_{i=1}^\infty |f_i| < \infty,$$

then it has norm

$$\| f \| = \sum_{i=1}^\infty |f_i|.$$

The inequality $\| f \| \le \sum_{i=1}^\infty |f_i|$ is obvious. On the other hand, if $\sum_{i=1}^\infty |f_i| = a$, then for every $\epsilon > 0$ it is possible to find an N such that $\sum_{i=1}^N |f_i| > a - \epsilon$.

We now set

$$x_n = \begin{cases} \left. \begin{cases} +1 \text{ if } f_n > 0 \\ -1 \text{ if } f_n < 0 \\ 0 \text{ if } f_n = 0 \end{cases} \right\} n \le N. \\ 0 \text{ if } n > N \end{cases}$$

Then

$$|f(x)| = \sum_{n=1}^N |f_n| > a - \epsilon,$$

whence it follows that $\| f \| = a$.

We shall prove that all functionals in the space c have the form (2). We shall set

$$e_n = (0, 0, \cdots, 1, 0, \cdots),$$

i.e. e_n denotes the sequence in which the n-th entry is unity and the remaining are zeros.

Let the functional $f(x)$ be given; we denote $f(e_n)$ by f_n. If

(3) $$x = (x_1, x_2, \cdots, x_n, 0, \cdots),$$

then

$$x = x_1 e_1 + x_2 e_2 + \cdots + x_n e_n \quad \text{and} \quad f(x) = \sum_{i=1}^{n} f_i x_i.$$

The sum $\sum_{n=1}^{\infty} |f_n|$ is $< \infty$ for every bounded linear functional. If $\sum_{n=1}^{\infty} |f_n|$ were $= \infty$, then for every H it would be possible to find an N such that

$$\sum_{n=1}^{N} |f_n| > H.$$

We construct the element x in the following way:

$$x_n = \begin{cases} 1 \text{ if } f_n > 0 \\ -1 \text{ if } f_n < 0 \\ 0 \text{ if } f = 0 \end{cases} n \leq N. \\ 0 \text{ if } n > N$$

The norm of such an element is equal to unity, and

$$|f(x)| = \sum_{i=1}^{N} f_i x_i = \sum_{i=1}^{N} |f_i| > H = H \| x \|,$$

which contradicts the assumption concerning the boundedness of the functional.

The set of elements of the form (3) is everywhere dense in the space c. Therefore the continuous linear functional is uniquely defined by its values on this set. Thus, for every x

$$f(x) = \sum_{n=1}^{\infty} f_i x_i.$$

The space which is conjugate to the space c consists of sequences $(f_1, f_2, \cdots, f_n, \cdots)$ satisfying the condition $\sum_{i=1}^{\infty} |f_n| < \infty$.

3. Let the space consist of sequences

$$x = (x_1, x_2, \cdots, x_n, \cdots), \qquad \sum_{i=1}^{\infty} |x_i| < \infty$$

with norm $\| x \| = \sum_{n=1}^{\infty} |x_n|$.

It can be proved that the space conjugate to this space is the space of bounded sequences

$$f = (f_1, f_2, \cdots, f_n, \cdots)$$

with norm $\| f \| = \sup_n |f_n|$.

In all the examples of finite-dimensional spaces introduced above, the space which is conjugate to the conjugate space coincides with the initial space. This is always so in the finite-dimensional case. However, as Examples 2 and 3 show, in the infinite-dimensional case the space conjugate to the conjugate space may not coincide with the initial space.

We consider cases when this coincidence holds also in infinite-dimensional space.

4. The space l_2 consists of sequences

$$x = (x_1, x_2, \cdots, x_n, \cdots),$$

with $\sum_{i=1}^{\infty} x_i^2 < \infty$ and norm $\| x \| = (\sum_{i=1}^{\infty} x_i^2)^{\frac{1}{2}}$. All functionals in the space l_2 have the form

$$f(x) = \sum_{i=1}^{\infty} f_i x_i.$$

We shall prove this assertion.

To each functional there is set into correspondence the sequence $f_1, f_2, \cdots, f_n, \cdots$ of its values on the elements $e_1, e_2, \cdots, e_n, \cdots$ defined exactly as in Example 2, above.

If the functional is bounded, then $\sum_{i=1}^{\infty} f_i^2 < \infty$. We shall assume the contrary, i.e. we shall assume that for every H there exists an N such that

$$\sum_{i=1}^{N} f_i^2 = U \geq H.$$

If we apply the functional under consideration to the element

$$x = (f_1, f_2, \cdots, f_N, 0, \cdots), \qquad \| x \| = U^{\frac{1}{2}},$$

we obtain

$$f(x) = \sum_{i=1}^{N} f_i^2 = U \geq H^{\frac{1}{2}} \| x \|,$$

contrary to the assumption that the functional is bounded.

Since the functional f is linear, its values on the elements of the form $x = (x_1, x_2, \cdots, x_n, 0, \cdots)$ are easily found; on all other elements of the space the values of f are found from continuity considerations and we always have

$$f(x) = \sum_{i=1}^{\infty} f_i x_i.$$

The norm of the functional f equals $(\sum_{i=1}^{\infty} f_i^2)^{\frac{1}{2}}$. This is established with the aid of the Schwarz inequality.

5. The space l_p is the space of all sequences of the form

$$x = (x_1, x_2, \cdots, x_n, \cdots), (\sum_{i=1}^{\infty} x_i^p)^{1/p} < \infty, \| x \| = (\sum_{i=1}^{\infty} x_i^p)^{1/p}.$$

The conjugate space of l_p is the space l_q, where $1/p + 1/q = 1$. The proof is analogous to the preceding proof. *Hint*: Use Hölder's inequality.

§25. Extension of linear functionals

THEOREM (HAHN-BANACH). *Every linear functional $f(x)$ defined on a linear subspace G of a normed linear space E can be extended to the entire* *space with preservation of norm, i.e. it is possible to construct a linear functional $F(x)$ such that*

1) $F(x) = f(x), \quad x \in G,$

2) $\| F \|_E = \| f \|_G .$

Proof. The theorem will be proved for a separable space E, although in actuality it is valid also in spaces which are not separable.

First, we shall extend the functional to the linear subspace G_1 obtained by adding to G some element $x_0 \notin G$. An arbitrary element of this subspace is uniquely representable in the form

$$y = tx_0 + x, \quad x \in G.$$

If the functional sought exists, then

$$F(y) = tF(x_0) + f(x)$$

or, if we set $-F(x_0) = c$, then $F(y) = f(x) - ct$.

In order that the norm of the functional be not increased when it is continued it is necessary to find a c such that the inequality

$$(1) \qquad | f(x) - ct | \le \| f \| \, \| x + tx_0 \|$$

be fulfilled for all $x \in G$.

If we denote the element x/t by z ($z \in G$), the inequality (1) can be rewritten

$$| f(z) - c | \le \| f \| \, \| z + x_0 \|.$$

This inequality is equivalent to the following inequalities:

$$-\| f \| \, \| z + x_0 \| \le f(z) - c \le \| f \| \, \| z + x_0 \|,$$

or, what amounts to the same thing,

$$f(z) + \| f \| \, \| z + x_0 \| \ge c \ge f(z) - \| f \| \, \| z + x_0 \|,$$

for all $z \in G$. We shall prove that such a number c always exists. To do this we shall show that for arbitrary elements z', $z'' \in G$ we always have

$$(2) \qquad f(z'') + \| f \| \, \| z'' + x_0 \| \ge f(z') - \| f \| \, \| z' + x_0 \|.$$

But this follows directly from the obvious inequality

$$f(z') - f(z'') \le \| f \| \, \| z' - z'' \| = \| f \| \, \| z' + x_0 - (z'' + x_0) \|$$
$$\le \| f \| \, \| z' + x_0 \| + \| f \| \, \| z'' + x_0 \|.$$

We introduce the notation:

$$c' = \inf \{f(z) + \| f \| \, \| z + x_0 \|; z \in G\},$$
$$c'' = \sup \{f(z) - \| f \| \, \| z + x_0 \|; z \in G\}.$$

It follows from inequality (2) that $c'' \leq c'$.

We take an arbitrary c such that $c'' \leq c \leq c'$. We set

$$F_1(x) = f(x) - ct$$

for the elements of the subspace $G_1 = \{G \cup x_0\}$. We obtain the linear functional F_1, where $\| F_1 \| = \| f \|$.

The separable space E contains a denumerable everywhere dense set $x_1, x_2, \cdots, x_n, \cdots$. We shall construct the linear subspaces

$$G_1 = \{G \cup x_0\},$$
$$G_2 = \{G_1 \cup x_1\},$$
$$\cdots\cdots\cdots\cdots\cdots$$
$$G_{n+1} = \{G_n \cup x_n\},$$
$$\cdots\cdots\cdots\cdots\cdots$$

and define the functional F on them as follows: we construct functionals F_n which coincide with F_{n-1} on G_{n-1} and which have norm equal to $\| f \|$. Thus, we obtain the functional F defined on a set which is everywhere dense in E. At the remaining points of E the functional is defined by continuity: if $x = \lim_{n\to\infty} x_n$, then $F(x) = \lim_{n\to\infty} F(x_n)$. The inequality $| F(x) | \leq \| f \| \| x \|$ is valid since

$$| F(x) | = \lim_{n\to\infty} | F(x_n) | \leq \lim_{n\to\infty} \| f \| \| x_n \| = \| f \| \| x \|.$$

This completes the proof of the theorem on the extension of a functional.

COROLLARY. Let x_0 be an arbitrary nonzero element in R and let M be an arbitrary positive number. Then there exists a linear functional $f(x)$ in R such that

$$\| f \| = M \quad \text{and} \quad f(x_0) = \| f \| \| x_0 \|.$$

In fact, if we set $f(tx_0) = tM \| x_0 \|$, we obtain a linear functional with norm equal to M which is defined on the one-dimensional subspace of elements of the form tx_0 and then, by the Hahn-Banach theorem, we can extend it to all of R without increasing the norm. The geometric interpretation of this fact is the following: in a Banach space through every point x_0 there can be drawn a hyperplane which is tangent to the sphere $\| x \| = \| x_0 \|$.

§26. The second conjugate space

Inasmuch as the totality $\bar{R}$ of linear functionals on a normed linear space R itself represents a normed linear space it is possible to speak of the space $\bar{\bar{R}}$ of linear functionals on $\bar{R}$, i.e. of the second conjugate space with respect to R, and so forth. We note first of all that every element in R defines a linear functional in $\bar{R}$. In fact, let

$$\psi_{x_0}(f) = f(x_0),$$

where x_0 is a fixed element in R and f runs through all of $\bar{R}$. Thus, to each $f \in \bar{R}$ there is set into correspondence some number $\psi_{x_0}(f)$. In this connection we have

$$\psi_{x_0}(\alpha f_1 + \beta f_2) = \alpha f_1(x_0) + \beta f_2(x_0) = \alpha \psi_{x_0}(f_1) + \beta \psi_{x_0}(f_2)$$

and

$$|\psi_{x_0}(f)| \leq \|f\| \|x_0\| \qquad \text{(boundedness)},$$

i.e. $\psi_{x_0}(f)$ is a bounded linear functional on $\bar{R}$.

Besides the notation $f(x)$ we shall also use the more symmetric notation:

$$(1) \qquad\qquad\qquad (f, x)$$

which is analogous to the symbol used for the scalar product. For fixed $f \in \bar{R}$ we can consider this expression as a functional on R and for fixed $x \in R$ as a functional on $\bar{R}$.

From this it follows that the norm of every $x \in R$ is defined in two ways: firstly, its norm is defined as an element in R, and secondly, as the norm of a linear functional on $\bar{R}$, i.e. as an element in $\bar{\bar{R}}$. Let $\|x\|$ denote the norm of x taken as an element in R and let $\|x\|_2$ be the norm of x taken as an element in $\bar{\bar{R}}$. We shall show that in fact $\|x\| = \|x\|_2$. Let f be an arbitrary nonzero element in $\bar{R}$. Then

$$|(f, x)| \leq \|f\| \|x\|, \qquad \|x\| \geq |(f, x)|/\|f\|;$$

since the left member of the last inequality does not depend on f, we have

$$\|x\| \geq \sup \{|(f, x)|/\|f\|; f \in \bar{R}, f \neq 0\} = \|x\|_2.$$

But, according to the corollary to the Hahn-Banach theorem, for every $x \in R$ a linear functional f_0 can be found such that

$$|(f_0, x)| = \|f_0\| \|x\|.$$

Consequently,

$$\sup \{|(f, x)|/\|f\|; f \in R\} = \|x\|,$$

i.e. $\|x\|_2 = \|x\|$.

This proves the following theorem.

THEOREM. *R is isometric to some linear manifold in* $\bar{\bar{R}}$.

Inasmuch as we agreed not to distinguish between isometric spaces this theorem can be formulated as follows: $R \subset \bar{\bar{R}}$.

The space R is said to be *reflexive* in case $\bar{\bar{R}} = R$. If $\bar{\bar{R}} \neq R$, then R is said to be *irreflexive*.

Finite-dimensional space R^n and the space l_2 are examples of reflexive spaces (we even have $\bar{R} = R$ for these spaces).

The space c of all sequences which converge to zero is an example of a complete irreflexive space. In fact, above (§24, Examples 2 and 3) we proved that the conjugate space of the space c is the space l of numerical sequences $(x_1, x_2, \cdots, x_n, \cdots)$ which satisfy the condition $\sum_{n=1}^{\infty} |x_n| < \infty$, to which in turn the space m of all bounded sequences is conjugate. The spaces c and m are not isometric. This follows from the fact that c is separable and m is not. Thus, c is irreflexive.

The space $C[a, b]$ of continuous functions on a closed interval $[a, b]$ is also irreflexive. However, we shall not stop to prove this assertion. (The following stronger assertion can also be proved: *No* normed linear space exists for which $C[a, b]$ is the conjugate space.)

A. I. Plessner established that for an arbitrary normed space R there exist only two possibilities: either the space R is reflexive, i.e. $R = \bar{\bar{R}} = \bar{\bar{\bar{\bar{R}}}} = \cdots$; $\bar{R} = \bar{\bar{\bar{R}}} = \cdots$; or the spaces $R, \bar{R}, \bar{\bar{R}}, \cdots$ are all distinct.

The space $l_p (p > 1)$ is an example of a reflexive space (since $\bar{l}_p = l_q$, where $1/p + 1/q = 1$, we have $\bar{\bar{l}}_p = \bar{l}_q = l_p$).

§27. Weak convergence

The concept of so-called weak convergence of elements in a normed linear space plays an important role in many questions of analysis.

DEFINITION. A sequence $\{x_n\}$ of elements in a normed linear space R *converges weakly* to the element x if

1) The norms of the elements x_n are uniformly bounded: $\|x_n\| \leq M$, and

2) $f(x_n) \to f(x)$ for every $f \in \bar{R}$.

(It can be shown that Condition 1 follows from 2; we shall not carry out this proof.)

Condition 2 can be weakened slightly; namely, the following theorem is true.

THEOREM 1. *The sequence* $\{x_n\}$ *converges weakly to the element* x *if*

1) $\|x_n\| \leq M$, *and*

2) $f(x_n) \to f(x)$ *for every* $f \in \Delta$, *where* Δ *is a set whose linear hull is everywhere dense in* $\bar{R}$.

Proof. It follows from the conditions of the theorem and the definition of a linear functional that if φ is a linear combination of elements in Δ then $\varphi(x_n) \to \varphi(x)$.

Now let φ be an arbitrary linear functional on R and let $\{\varphi_k\}$ be a sequence of functionals which converges to φ, each of which is a linear combination of elements in Δ. We shall show that $\varphi(x_n) \to \varphi(x)$. Let M be such that $\|x_n\| \leq M$ $(n = 1, 2, \cdots)$ and $\|x\| \leq M$.

Let us evaluate the difference $|\varphi(x_n) - \varphi(x)|$. Since $\varphi_k \to \varphi$, given an arbitrary $\epsilon > 0$ a K can be found such that for all $k > K$,

$$\|\varphi - \varphi_k\| < \epsilon;$$

it follows from this that

$$| \varphi(x_n) - \varphi(x) | \leq | \varphi(x_n) - \varphi_k(x_n) | + | \varphi_k(x_n) - \varphi_k(x) |$$
$$+ | \varphi_k(x) - \varphi(x) | \leq \epsilon M + \epsilon M + | \varphi_k(x_n) - \varphi_k(x) |.$$

But by assumption $| \varphi_k(x_n) - \varphi_k(x) | \to 0$ as $n \to \infty$. Consequently, $| \varphi(x_n) - \varphi(x) | \to 0$ as $n \to \infty$.

If the sequence $\{x_n\}$ converges in norm to x, that is, if $\| x_n - x \| \to 0$ as $n \to \infty$, then such convergence is frequently called strong convergence to distinguish it from weak convergence.

If a sequence $\{x_n\}$ converges strongly to x, it also converges weakly to the same limit. In fact, if $\| x_n - x \| \to 0$, then

$$| f(x_n) - f(x) | \leq \| f \| \| x_n - x \| \to 0$$

for an arbitrary linear functional f. The converse is not true in general: strong convergence does not follow from weak convergence. For example, in l_2 the sequence of vectors

$$e_1 = (1, 0, 0, \cdots),$$
$$e_2 = (0, 1, 0, \cdots),$$
$$e_3 = (0, 0, 1, \cdots),$$
$$\cdots \cdots \cdots \cdots \cdots$$

converges weakly to zero. In fact, every linear functional f in l_2 can be represented as the scalar product with some fixed vector

$$a = (a_1, a_2, \cdots, a_n, \cdots), \qquad f(x) = (x, a);$$

hence,

$$f(e_n) = (e_n, a) = a_n.$$

Since $a_n \to 0$ as $n \to \infty$ for every $a \in l_2$, we have $\lim f(e_n) = 0$ for every linear functional in l_2.

But at the same time the sequence $\{e_n\}$ does not converge in the strong sense to any limit.

We shall investigate what weak convergence amounts to in several concrete spaces.

EXAMPLES. 1. In *finite-dimensional space* R^n weak and strong convergence coincide. In fact, consider functionals corresponding to multiplication by the elements

$$e_1 = (1, 0, 0, \cdots, 0),$$
$$e_2 = (0, 1, 0, \cdots, 0),$$
$$\cdots \cdots \cdots \cdots \cdots \cdots$$
$$e_n = (0, 0, 0, \cdots, 1).$$

If $\{x_k\}$ converges weakly to x, then

$$(x_k, e_i) = x_k^{(i)} \to x^{(i)} \qquad (i = 1, 2, \cdots, n),$$

i.e. the first coordinates of the vectors x_k tend to the first coordinate of the vector x, their second coordinates tend to the second coordinate of the vector x, and so forth. But then

$$\rho(x_k, x) = \{\textstyle\sum_{i=1}^{n} (x_k^{(i)} - x^{(i)})^2\}^{\frac{1}{2}} \to 0,$$

i.e. $\{x_k\}$ converges strongly to x. Since strong convergence always implies weak convergence, our assertion is proved.

2. *Weak convergence in l_2*. Here we can take for the set Δ, linear combinations of whose elements are everywhere dense in l_2, the totality of vectors

$$e_1 = (1, 0, 0, \cdots),$$
$$e_2 = (0, 1, 0, \cdots),$$
$$e_3 = (0, 0, 1, \cdots),$$
$$\cdots\cdots\cdots\cdots\cdots$$

If $x = (\xi_1, \xi_2, \cdots, \xi_n, \cdots)$ is an arbitrary vector in l_2, then the values assumed at x by the corresponding linear functionals are equal to $(x, e_n) = \xi_n$, i.e. to the coordinates of the vector x. Consequently, weak convergence of the sequence $\{x_n\}$ in l_2 means that the numerical sequence of the k-th coordinates of these vectors ($k = 1, 2, \cdots$) converges. We saw above that this convergence does not coincide with strong convergence in l_2.

3. *Weak convergence in the space of continuous functions*. Let $C[a, b]$ be the space of continuous functions defined on the closed interval $[a, b]$. It can be shown that the totality Δ of all linear functionals δ_{t_0}, each of which is defined as the value of the function at some fixed point t_0 (see Example 4, §23) satisfies the conditions of Theorem 1, i.e. linear combinations of these functionals are everywhere dense in $\bar{C}[a, b]$. For each such functional δ_{t_0}, the condition $\delta_{t_0} x_n(t) \to \delta_{t_0} x(t)$ is equivalent to the condition $x_n(t_0) \to x(t_0)$.

Thus, weak convergence of a sequence of continuous functions means that this sequence is a) uniformly bounded and b) convergent at every point.

It is clear that this convergence does not coincide with convergence in norm in $C[a, b]$, i.e. it does not coincide with uniform convergence of continuous functions. (Give a suitable example!)

§28. Weak convergence of linear functionals

We can introduce the concept of weak convergence of linear functionals as analogous to the concept of weak convergence of elements of a normed linear space R.

DEFINITION. A sequence $\{f_n\}$ of linear functionals converges weakly to the linear functional f if

1) $\| f_n \|$ are uniformly bounded; i.e. $\| f_n \| \leq M$, and
2) $f_n(x) \to f(x)$ for every element $x \in R$.

(This is usually called weak* convergence.—Trans.)

Weak convergence of linear functionals possesses properties which are analogous to properties stated above of weak convergence of elements, namely strong convergence (i.e. convergence in norm) of linear functionals implies their weak convergence, and it is sufficient to require the fulfillment of the condition $f_n(x) \to f(x)$ not of all $x \in R$, but only for a set of elements linear combinations of which are everywhere dense in R.

We shall consider one important example of weak convergence of linear functionals. Above (§23, Example 4), we spoke of the fact that the "δ-function", i.e. the functional on $C[a, b]$, which assigns to every continuous function its value at the point zero, can "in some sense" be considered as the limit of "ordinary" functions, each of which assumes the value zero outside some small neighborhood of zero and has an integral equal to 1. [We assume that the point $t = 0$ belongs to the interval (a, b). Of course, one can take any other point instead of $t = 0$.] Now we can state this assertion precisely. Let $\{\varphi_n(t)\}$ be a sequence of continuous functions satisfying the following conditions:

(1) 1) $\varphi_n(t) = 0$ for $|t| > 1/n$, $\varphi_n(t) \geq 0$,

   2) $\displaystyle\int_a^b \varphi_n(t)\, dt = 1$.

Then for an arbitrary continuous function $x(t)$ defined on the closed interval $[a, b]$, we have

$$\int_a^b \varphi_n(t)x(t)\, dt = \int_{-1/n}^{1/n} \varphi_n(t)x(t)\, dt \to x(0) \text{ as } n \to \infty.$$

In fact, by the mean-value theorem,

$$\int_{-1/n}^{1/n} \varphi_n(t)x(t)\, dt = x(\xi_n) \int_{-1/n}^{1/n} \varphi_n(t)\, dt = x(\xi_n), \qquad -1/n \leq \xi_n \leq 1/n;$$

when $n \to \infty$, $\xi_n \to 0$ and $x(\xi_n) \to x(0)$.

The expression

$$\int_a^b \varphi_n(t)x(t)\, dt$$

represents a linear functional on the space of continuous functions. Thus, the result we obtained can be formulated as follows: the δ-function is the limit of the sequence (1) in the sense of weak convergence of linear functionals.

The following theorem plays an important role in various applications of the concept of weak convergence of linear functionals.

THEOREM 1. *If the normed linear space R is separable, then an arbitrary bounded sequence of linear functionals on R contains a weakly convergent subsequence.*

Proof. Choose in R a denumerable everywhere dense set

$$\{x_1, x_2, \cdots, x_n, \cdots\}.$$

If $\{f_n\}$ is a bounded (in norm) sequence of linear functionals on R, then

$$f_1(x_1), f_2(x_1), \cdots, f_n(x_1), \cdots$$

is a bounded numerical sequence. Therefore we can select from $\{f_n\}$ a subsequence

$$f_1', f_2', \cdots, f_n', \cdots$$

such that the numerical sequence

$$f_1'(x_1), f_2'(x_1), \cdots$$

converges. Furthermore, from the subsequence $\{f_n'\}$ we can select a subsequence

$$f_1'', f_2'', \cdots, f_n'', \cdots$$

such that

$$f_1''(x_2), f_2''(x_2), \cdots$$

converges, and so forth. Thus, we obtain a system of sequences

$$f_1', f_2', \cdots, f_n', \cdots,$$
$$f_1'', f_2'', \cdots, f_n'', \cdots,$$
$$\cdots\cdots\cdots\cdots\cdots\cdots\cdots$$

each of which is a subsequence of the one preceding. Then taking the "diagonal" subsequence $f_1', f_2'', f_3''', \cdots$, we obtain a sequence of linear functionals such that $f_1'(x_n), f_2''(x_n), \cdots$ converges for all n. But then $f_1'(x)$, $f_2''(x), \cdots$ also converges for arbitrary $x \in R$. This completes the proof of the theorem.

The last theorem suggests the following question. Is it possible in the space $\bar{R}$, conjugate to a separable space, to introduce a metric so that the bounded subsets of the space $\bar{R}$ become compact with respect to this new metric? In other words, is it possible to introduce a metric in $\bar{R}$ so that convergence in the sense of this metric in $\bar{R}$ coincides with weak convergence of elements in $\bar{R}$ considered as linear functionals. Such a metric can in fact be introduced in $\bar{R}$.

Let $\{x_n\}$ be a denumerable everywhere dense set in R. Set

$$(2) \qquad \rho(f_1, f_2) = \sum_{n=1}^{\infty} |f_1(x_n) - f_2(x_n)|/2^n \|x_n\|$$

for any two elements f_1, $f_2 \in \bar{R}$. This series converges since its n-th term does not exceed $(\|f_1\| + \|f_2\|)/2^n$. The quantity (2) possesses all the properties of a distance. In fact, the first two axioms are obviously satisfied. We shall verify the triangle axiom.

Since

$$|f_1(x_n) - f_3(x_n)| = |f_1(x_n) - f_2(x_n) + f_2(x_n) - f_3(x_n)|$$
$$\leq |f_1(x_n) - f_2(x_n)| + |f_2(x_n) - f_3(x_n)|,$$

we have

$$\rho(f_1, f_3) \leq \rho(f_1, f_2) + \rho(f_2, f_3).$$

Direct verification shows that convergence in the sense of this metric is in fact equivalent to weak convergence in $\bar{R}$.

Now Theorem 1 can be formulated in the following manner.

THEOREM 1'. *In a space $\bar{R}$, which is the conjugate of a separable space, with metric (2), every bounded subset is compact.*

§29. Linear operators

1. *Definition of a linear operator. Boundedness and continuity.*

Let R and R' be two Banach spaces whose elements are denoted respectively by x and y. Let a rule be given according to which to each x in some set $X \subseteq R$ there is assigned some element y in the space R'. Then we say that an *operator* $y = Ax$ with range of values in R' has been defined on the set X.

DEFINITION 1. An operator A is said to be *linear* if the equality

$$A(\alpha_1 x_1 + \alpha_2 x_2) = \alpha_1 A x_1 + \alpha_2 A x_2$$

is satisfied for any two elements x_1, $x_2 \in X$ and arbitrary real numbers α_1, α_2.

DEFINITION 2. An operator A is said to be *bounded* if there exists a constant M such that

$$\|Ax\| \leq M \|x\|$$

for all $x \in X$.

DEFINITON 3. An operator A is said to be *continuous* if for arbitrary $\epsilon > 0$ there exists a number $\delta > 0$ such that the inequality

$$\|x' - x''\|_R < \delta \qquad (x', x'' \in X)$$

implies that

$$\|Ax' - Ax''\|_{R'} < \epsilon.$$

Only linear operators will be considered in the sequel. If the space R' is the real line, the operator $y = A(x)$ is a functional, and the formulated definitions of linearity, continuity and boundedness go over into the corresponding definitions introduced in §23 for functionals.

The following theorem is a generalization of Theorem 1, §23.

THEOREM 1. *Continuity is equivalent to boundedness for a linear operator.*

Proof. 1. Assume the operator A is bounded. The inequality $\| x' - x'' \| < \delta$ implies that

$$(1) \qquad \| Ax' - Ax'' \| = \| A(x' - x'') \| \leq M \| x' - x'' \| \leq M\delta,$$

where M is the constant occurring in the definition of boundedness. If we take $\delta < \epsilon/M$, inequality (1) yields $\| Ax' - Ax'' \| < \epsilon$, i.e. the operator A is continuous.

2. Assume now that the operator A is continuous. We shall prove that A is bounded by contradiction. We assume that A is not bounded. Then there exists a sequence

$$(2) \qquad x_1, x_2, \cdots, x_n, \cdots$$

such that

$$\| Ax_n \| > n \| x_n \|.$$

Set $z_n = x_n/n \| x_n \|$; it is obvious that $\| z_n \| = 1/n$, i.e. $z_n \to 0$ as $n \to \infty$. Consider the sequence

$$Az_n = Ax_n/n \| x_n \|$$

which is the map of the sequence $\{z_n\}$ under A. The norm of each element Az_n is not less than 1:

$$\| Az_n \| = \| Ax_n \|/n \| x_n \| \geq n \| x_n \|/ n\| x_n \| = 1.$$

Since for every linear operator, $A(0) = 0$, and $\lim_{n \to \infty} z_n = 0$, we obtain a contradiction of our hypothesis that the operator is continuous. Consequently, the operator A must be bounded.

EXAMPLE. *The general form of a linear operator mapping a finite-dimensional space into a finite-dimensional space.* Given an n-dimensional space R^n with basis $e_1, e_2, \cdots, e_n$, every point of this space can be represented in the form $x = \sum_{i=1}^{n} x_i e_i$.

A linear operator A maps R^n into the finite-dimensional space R^m with the basis $e_1', e_2', \cdots, e_m'$.

Let us consider the representation with respect to this basis of the images of the basis vectors of the space R^n:

$$Ae_i = \sum_{j=1}^{m} a_{ij} e_j'.$$

Now let $y = Ax$,

$$y = Ax = \sum_{i=1}^{n} x_i Ae_i = \sum_{i=1}^{n} x_i \sum_{j=1}^{m} a_{ij} e_j' = \sum_{j=1}^{m} d_j e_j',$$

where

$$(3) \qquad d_j = \sum_{i=1}^{n} a_{ij} x_i .$$

It is clear from formula (3) that to determine the operator A it is sufficient to give the coefficient matrix with entries a_{ij}.

A linear operator cannot map a finite-dimensional space into a space of greater dimension since all linear relations among the elements are preserved for their images.

2. *Norm of an operator. Sum and product of operators. Product of an operator by a scalar.*

DEFINITION 4. Let A be a bounded linear operator. This means that there exist numbers M such that

$$(4) \qquad \| Ax \| < M \| x \|$$

for all $x \in X$. The *norm* $\| A \|$ of the operator A is the greatest lower bound of the numbers M which satisfy condition (4). It follows from the definition of the norm of an operator that $\| Ax \| \leq \| A \| \| x \|$. But if $M < \| A \|$, then there exists an element x such that $\| Ax \| > M \| x \|$.

THEOREM 2. *If A is an arbitrary linear operator, then*

$$\| A \| = \sup \{ \| Ax \|; \| x \| = 1 \} = \sup \{ \| Ax \|/\| x \|; \| x \| \neq 0 \}.$$

Proof. Introduce the notation

$$\alpha = \sup \{ \| Ax \|; \| x \| = 1 \} = \sup \{ \| Ax \|/\| x \|; \| x \| \neq 0 \}.$$

We shall first prove that $\| A \| \geq \alpha$. Since $\alpha = \sup \{ \| Ax \|/\| x \|; \| x \| \neq 0 \}$, for arbitrary $\epsilon > 0$ there exists an element x_1 not zero such that $\| Ax_1 \|/\| x_1 \| > \alpha - \epsilon$ or $\| Ax_1 \| > (\alpha - \epsilon) \| x_1 \|$, which implies that $\alpha - \epsilon < \| A \|$ and hence that $\| A \| \geq \alpha$ because ϵ is arbitrary.

The inequality cannot hold. In fact, if we let $\| A \| - \alpha = \epsilon$, then $\alpha < \| A \| - \epsilon/2$. But this implies that the following inequalities hold for an arbitrary point x:

$$\| Ax \|/\| x \| \leq \alpha < \| A \| - \epsilon/2,$$

or

$$\| Ax \| \leq (\| A \| - \epsilon/2) \| x \|,$$

i.e. $\| A \|$ is not the greatest lower bound of those M for which $\| Ax \| \leq M \| x \|$. It is clear from this contradiction that $\| A \| = \alpha$.

In the sequel we shall make use of the above expression for the norm of an operator as equivalent to the original definition of the norm.

DEFINITION 5. Let A_1 and A_2 be two given continuous linear operators which transform the Banach space E into the Banach space E_1. The *sum* of these two operators is the operator A which puts the element $y \in E_1$ defined by the formula $y = A_1x + A_2x$ into correspondence with the element $x \in E$. It is easy to verify that $A = A_1 + A_2$ is also a linear operator.

THEOREM 3. *The following relation holds for the norms of the operators* A_1, A_2 *and* $A = A_1 + A_2$:

(5) $$\| A \| \leq \| A_1 \| + \| A_2 \|.$$

Proof. It is clear that

$$\| Ax \| = \| A_1x + A_2x \| \leq \| A_1x \| + \| A_2x \| \leq (\| A_1 \| + \| A_2 \|) \| x \|,$$

whence inequality (5) follows.

DEFINITION 6. Let A_1 and A_2 be continuous linear operators where A_1 transforms the Banach space E into the Banach space E_1 and A_2 transforms the Banach space E_1 into the Banach space E_2. The *product* of the operators A_1 and A_2 (denoted by $A = A_2A_1$) is the operator which sets the element $z \in E_2$ into correspondence with the element $x \in E$, where

$$z = A_2(A_1x).$$

THEOREM 4. *If* $A = A_2A_1$, *then*

(6) $$\| A \| \leq \| A_2 \| \| A_1 \|.$$

Proof. $\| Ax \| = \| A_2(A_1x) \| \leq \| A_2 \| \| A_1x \| \leq \| A_2 \| \| A_1 \| \| x \|,$ whence the assertion of the theorem follows.

The sum and product of three or more operators are defined by iteration. Both operations are associative.

The product of the operator A and the real number k (denoted by kA) is defined in the following manner: the operator kA puts the element $k(Ax)$ of the space E_1 into correspondence with the element $x \in E$.

It is easy to verify that with respect to the operations of addition and multiplication by a scalar introduced above the bounded linear operators form a linear space. If we introduce the norm of the operator in the way indicated above we can form a *normed* linear space.

EXERCISE. Prove that the space of bounded linear operators which transform the space E_1 into a complete space E_2 is complete.

3. *The inverse operator.*

Let us consider the operator T which transforms the Banach space E into the Banach space E_1 :

$$Tx = y, \qquad x \in E, \qquad y \in E_1.$$

DEFINITION 7. The operator T is said to have an *inverse* if for every $y \in E_1$ the equation

(7) $$Tx = y$$

has a unique solution.

To each $y \in E_1$ we can put into correspondence the solution of equation (7). The operator which realizes this correspondence is said to be the *inverse* of T and is denoted by T^{-1}.

THEOREM 5. *The operator T^{-1} which is the inverse of the linear operator T is also linear.*

Proof. To prove the linearity of T^{-1} it is sufficient to verify that the equality

$$T^{-1}(\alpha_1 y_1 + \alpha_2 y_2) = \alpha_1 T^{-1} y_1 + \alpha_2 T^{-1} y_2$$

is valid. Denote Tx_1 by y_1 and Tx_2 by y_2. Since T is linear, we have

$$(8) \qquad\qquad T(\alpha_1 x_1 + d_2 x_2) = \alpha_1 y_1 + \alpha_2 y_2.$$

By the definition of the inverse operator,

$$T^{-1} y_1 = x_1, \qquad T^{-1} y_2 = x_2;$$

whence, multiplying these equations by α_1 and α_2 respectively, and adding, we obtain:

$$\alpha_1 T^{-1} y_1 + \alpha_2 T^{-1} y_2 = \alpha_1 x_1 + \alpha_2 x_2.$$

On the other hand, from (8) and from the definition of the inverse operator it follows that

$$\alpha_1 x_1 + \alpha_2 x_2 = T^{-1}(\alpha_1 y_1 + \alpha_2 y_2)$$

which together with the preceding equalities yields

$$T^{-1}(\alpha_1 y_1 + \alpha_2 y_2) = \alpha_1 T^{-1} y_1 + \alpha_2 T^{-1} y_2.$$

THEOREM 6. *If T is a bounded linear operator whose inverse T^{-1} exists, then T^{-1} is bounded.*

We shall need the following two lemmas in the proof of this theorem.

LEMMA 1. *Let M be an everywhere dense set in the Banach space E. Then an arbitrary element $y \in E$, $y \neq 0$, can be developed in the series*

$$y = y_1 + y_2 + \cdots + y_n + \cdots,$$

where $y_k \in M$ and $\| y_k \| \leq 3 \| y \|/2^k$.

Proof. We construct the sequence of elements y_k in the following way: we choose y_1 so that

$$(9) \qquad\qquad \| y - y_1 \| \leq \| y \|/2.$$

This is possible because inequality (9) defines a sphere of radius $\| y \|/2$ with center at the point y, whose interior must contain an element of M (since M is everywhere dense in E). We choose $y_2 \in M$ such that $\| y - y_1 - y_2 \| \leq \| y \|/4$, y_3 such that $\| y - y_1 - y_2 - y_3 \| \leq \| y \|/8$, and in general, y_n such that $\| y - y_1 - \cdots - y_n \| \leq \| y \|/2^n$.

Such a choice is always possible because M is everywhere dense in E. By construction of the elements y_k,

$$\| y - \sum_{k=1}^{n} y_k \| \to 0 \quad \text{as} \quad n \to \infty,$$

i.e. the series $\sum_{k=1}^{\infty} y_k$ converges to y.

To evaluate the norms of the elements y_k we proceed as follows:

$$\| y_1 \| = \| y_1 - y + y \| \leq \| y_1 - y \| + \| y \| = 3 \| y \|/2,$$

$$\| y_2 \| = \| y_2 + y_1 - y - y_1 + y \|$$
$$\leq \| y - y_1 - y_2 \| + \| y - y_1 \| \leq 3 \| y \|/4.$$

Finally, we obtain

$$\| y_n \| = \| y_n + y_{n-1} + \cdots + y_1 - y + y - y_1 - \cdots - y_{n-1} \|$$
$$\leq \| y - y_1 - \cdots - y_n \| + \| y - y_1 - \cdots - y_{n-1} \| = 3 \| y \|/2^n.$$

This proves Lemma 1.

LEMMA 2. *If the Banach space E is the sum of a denumerable number of sets: $E = \bigcup_{n=1}^{\infty} M_n$, then at least one of these sets is dense in some sphere.*

Proof. Without loss of generality, we can assume that

$$M_1 \subseteq M_2 \subseteq M_3 \subseteq \cdots .$$

We shall assume that all the sets M_i are nowhere dense, i.e. that in the interior of every sphere there exists another sphere which does not contain a single point of M_k, $k = 1, 2, \cdots$.

Take an arbitrary sphere S_0; in it there exists a sphere S_1 which does not contain a single point of M_1; in S_1 there exists a sphere S_2 which does not contain a single point of M_2; and so forth. We obtain a sequence of nested spheres which can be chosen so that the radius of the sphere S_n converges to zero as $n \to \infty$. In a Banach space such spheres have a common point. This point is an element of E but it does not belong to any of the sets M_n. This contradicts the hypothesis of the lemma and proves Lemma 2.

Proof of Theorem 6. In the space E_1 let us consider the sets M_k, where M_k is the set of all y for which the inequality $\| T^{-1}y \| \leq k \| y \|$ holds.

Every element of E_1 is contained in some M_k, i.e. $E_1 = \bigcup_{n=1}^{\infty} M_n$. By Lemma 2, at least one of the M_n is dense in some sphere S_0. In the interior of the sphere S_0 let us consider the spherical shell P consisting of the points z for which

$$\beta < \| z - y_0 \| < \alpha,$$

where

$$0 < \beta < \alpha, \quad y_0 \in M_n .$$

If we translate the spherical shell P so that its center coincides with the origin of coordinates, we obtain the spherical shell P_0.

We shall show that some set M_n is dense in P_0. Consider $z \in P$; then $z - y_0 \in P_0$. Furthermore, let $z \in M_n$. By virtue of the choice of z and y_0 we obtain:

$$\| T^{-1}(z - y_0) \| \leq \| T^{-1}z \| + \| T^{-1}y_0 \|$$

$$\leq n(\| z \| + \| y_0 \|) \leq n(\| z - y_0 \| + 2 \| y_0 \|)$$

$$= n \| z - y_0 \| [1 + 2 \| y_0 \|/\| z - y_0 \|] \leq n \| z - y_0 \| (1 + 2 \| y_0 \|/\beta).$$

The quantity $n(1 + 2 \| y_0 \|/\beta)$ does not depend on z. Denote $n(1 + 2 \| y_0 \|/\beta)$ by N. Then by definition $z - y_0 \in M_N$ and M_N is dense in P_0 because M_N is obtained from M_n, as was P_0 from P, by means of a translation by y_0 and M_n is dense in P. Consider an arbitrary element y in E_1. It is always possible to choose λ so that $\beta < \| \lambda y \| < \alpha$. For λy we can construct a sequence $y_k \in M_N$ which converges to λy. Then the sequence $(1/\lambda)y_k$ converges to y. (It is obvious that if $y_k \in M_N$, then $(1/\lambda)y_k \in M_N$ for arbitrary real $1/\lambda$.)

We have proved that for arbitrary $y \in E_1$ a sequence of elements of M_N can be found which converges to y, i.e. that M_N is everywhere dense in E_1.

Consider $y \in E_1$; by Lemma 1, y can be developed in a series of elements in M_N:

$$y = y_1 + y_2 + \cdots + y_n + \cdots,$$

where $\| y_n \| < 3 \| y \|/2^n$.

Consider in the space E the series formed from the inverse images of the y_k, i.e. from $x_k = T^{-1}y_k$:

$$\sum_{k=1}^{\infty} x_k = x_1 + x_2 + \cdots + x_n + \cdots.$$

This series converges to some element x since the following inequality holds: $\| x_n \| = \| T^{-1}y_n \| \leq N \| y_n \| < 3N \| y \|/2^n$ and consequently, $\| x \| \leq \sum_{k=1}^{\infty} \|x_k\| \leq 3N \| y \| \sum_{n=1}^{\infty} (\frac{1}{2})^n = 3 \| y \| N$.

By virtue of the convergence of the series $\sum_{n=1}^{\infty} x_n$ and the continuity of T we can apply T to the series. We obtain:

$$Tx = Tx_1 + Tx_2 + \cdots = y_1 + y_2 + \cdots = y,$$

whence $x = T^{-1}y$. We have

$$\| x \| = \| T^{-1}y \| \leq 3N \| y \|$$

and since this estimate is valid for arbitrary y, it follows that T^{-1} is bounded.

THEOREM 7. *An operator which closely approximates an operator whose inverse exists has an inverse, i.e. if T_0 is a linear operator which has an in-*

verse and which maps the space E into the space E_1, and ΔT is an operator which also maps E into E_1, where $\| \Delta T \| < 1/\| T_0^{-1} \|$, then the operator $T = T_0 + \Delta T$ has an inverse.

Proof. Let $y \in E_1$. We wish to find a unique $x_0 \in E$ such that

$$y = Tx_0 = T_0x_0 + \Delta Tx_0.$$

If we apply the operator T_0^{-1} to this equation, we obtain

(10) $$T_0^{-1}y = x_0 + T_0^{-1}\Delta Tx_0;$$

if we denote $T_0^{-1}y$ by $z \in E$ and $T_0^{-1}\Delta T$ by A, then equation (10) can be written in the form

$$z = x_0 + Ax_0,$$

where A is an operator which maps the Banach space E into itself and $\| A \| < 1$.

The mapping $x' = z - Ax$ is a contraction mapping of the space E into itself; consequently, it has a unique fixed point which is the unique solution of equation (10) and this means that the operator T has an inverse.

Theorem 8. *The operator which is the inverse of the operator $T = I - A$, where I is the identity operator and the operator A (of E into E) has norm less than 1 ($\| A \| < 1$), can be written in the form*

(11) $$(I - A)^{-1} = \sum_{k=0}^{\infty} A^k.$$

Proof. Consider the following transformation of the space E into itself:

$$y = Tx, \qquad x \in E, \qquad y \in E.$$

The mapping $x' = y + Ax$ is a contraction mapping of the space E into itself by virtue of the condition that $\| A \| < 1$.

We solve the equation $x = y + Ax$ by means of the iterations: $x_{n+1} = y + Ax_n$. If we set $x_0 = 0$, we obtain $x_1 = y; x_2 = y + Ay; x_3 = y + Ay + A^2y; \cdots; x_n = y + Ay + A^2y + \cdots + A^{n-1}y$.

As $n \to \infty$, x_n tends to the unique solution of the equation $x = y + Ax$, i.e. $x = \sum_{k=0}^{\infty} A^k y$, whence

$$(I - A)^{-1}y = \sum_{k=0}^{\infty} A^k y,$$

which yields equation (11).

4. *Adjoint operators.*

Consider the linear operator $y = Ax$ which maps the Banach space E into the Banach space E_1. Let $g(y)$ be a linear functional defined on E_1, i.e. $g(y) \in \bar{E}_1$. Apply the functional g to the element $y = Ax$; $g(Ax)$, as is easily verified, is a linear functional defined on E; denote it by $f(x)$. The functional $f(x)$ is thus an element of the space $\bar{E}$. We have assigned to each functional $g \in \bar{E}_1$ a functional $f \in \bar{E}$, i.e. we have obtained an operator which maps $\bar{E}_1$ into $\bar{E}$. This operator is called the *adjoint* operator of the operator A and is denoted by A^* or by $f = A^*g$.

If we use the notation (f, x) for the functional $f(x)$, we obtain $(g, Ax) = (f, x)$, or $(g, Ax) = (A^*g, x)$. This relation can be taken for the definition of the adjoint operator.

EXAMPLE. *The expression for the adjoint operator in finite-dimensional space.* Euclidean n-space E^n is mapped by the operator A into Euclidean m-space E^m. The operator A is given by the matrix (a_{ij}).

The mapping $y = Ax$ can be written in the form of the system

$$y_i = \sum_{j=1}^{n} a_{ij}x_j, \qquad i = 1, 2, \cdots, m.$$

The operator $f(x)$ can be written in the form

$$f(x) = \sum_{j=1}^{n} f_j x_j.$$

The equalities

$$f(x) = g(Ax) = \sum_{i=1}^{m} g_i y_i = \sum_{i=1}^{m} \sum_{j=1}^{n} g_i a_{ij} x_j = \sum_{j=1}^{n} x_j \sum_{i=1}^{m} g_i a_{ij}$$

imply that $f_j = \sum_{i=1}^{m} g_i a_{ij}$. Since $f = A^*g$, it follows that the operator A^* is given by the transpose of the matrix for the operator A.

We shall now list the basic properties of adjoint operators.

1. The adjoint operator of the sum of two linear operators is equal to the sum of the adjoint operators:

$$(A + B)^* = A^* + B^*.$$

Let $f_1 = A^*g$, $f_2 = B^*g$, or $f_1(x) = g(Ax)$, $f_2(x) = g(Bx)$; then $(f_1 + f_2)(x) = g(Ax + Bx) = g[(A + B)x]$, whence $(A + B)^* = A^* + B^*$.

2. The adjoint operator of the operator kA, where k is a scalar multiplier, is equal to the adjoint operator of A, multiplied by k:

$$(kA)^* = kA^*.$$

The verification of this property is elementary and is left to the reader.

3. $I^* = I$, i.e. the adjoint of the identity operator on E is the identity operator on $\bar{E}$.

THEOREM 9. *The operator A^*, the adjoint of a linear operator A which maps the Banach space E into the Banach space E_1, is also linear and* $\| A^* \| = \| A \|$.

Proof. The linearity of the operator A^* is obvious. We shall prove the equality of the norms. By virtue of the properties of the norm of an operator we have:

$$|f(x)| = |g(Ax)| \leq \| g \| \| Ax \| \leq \| g \| \| A \| \| x \|,$$

whence $\| f \| \leq \| A \| \, \| g \|$ or $\| A^*g \| \leq \| A \| \, \| g \|$ and consequently,

$$(12) \qquad\qquad \| A^* \| \leq \| A \|.$$

Let $x \in E$ and form $y_0 = Ax/\| Ax \| \in E_1$; it is clear that $\| y_0 \| = 1$. From the corollary to the Hahn-Banach theorem, there exists a functional g such that $\| g \| = 1$ and $g(y_0) = 1$, i.e. $g(Ax) = \| Ax \|$.

From the inequalities $\| Ax \| = | (g, Ax) | = | (A^*g, x) | \leq \| A^*g \| \, \| x \| \leq \| A^* \| \, \| g \| \, \| x \| = \| A^* \| \, \| x \|$ we obtain $\| A \| \leq \| A^* \|$ which, combined with inequality (12), yields $\| A \| = \| A^* \|$.

ADDENDUM TO CHAPTER III
Generalized Functions

In a number of cases in analysis and in its various applications, for example in theoretical physics, the need arises to introduce various "generalized" functions in addition to the "ordinary" functions. A typical example of this is the well-known δ-function which we have already mentioned above (§23, Example 4).

We wish to emphasize, however, that these concepts, which are discussed briefly in this addendum, did not in any sense originate in an attempt to generalize the concepts of analysis merely for the sake of generalizing. Rather, they were suggested by perfectly concrete problems. Moreover, essentially the same concepts were used by physicists for quite some time before they attracted the attention of mathematicians.

The method of introducing generalized functions which we shall use below originated in the work of S. L. Sobolev, published in 1935–36. Later, these ideas were developed in a somewhat extended form by L. Schwartz.

Consider on the real line the set D of functions $\varphi(x)$ each of which vanishes outside some interval (where for each φ there is a corresponding interval) and has derivatives of all orders. The elements of D can be added and multiplied by a scalar in the usual way. Thus, D is a linear space. We shall not introduce a norm into this space; however, in D one can define in a natural way the convergence of a sequence of elements. We shall say that $\varphi_n \to \varphi$ if: 1) there exists an interval in the exterior of which all φ_n and φ are equal to zero and 2) the sequence of derivatives $\varphi_n^{(k)}$ of order k ($k = 0, 1, 2, \cdots$) (where the derivative of order zero is understood to be as usual the function itself) converges uniformly to $\varphi^{(k)}$ on this interval. The fact that this concept of convergence is not connected with any norm does not give rise to any inconveniences.

We now introduce the concept of generalized function.

DEFINITION 1. *A generalized function* (with values on the real line $-\infty < t < \infty$) is any linear functional $T(\varphi)$ defined on the space D. Thus, $T(\varphi)$ satisfies the following conditions:

1. $T(\alpha\varphi_1 + \beta\varphi_2) = \alpha T(\varphi_1) + \beta T(\varphi_2)$;
2. If $\varphi_n \to \varphi$ (in the sense indicated above), then $T(\varphi_n) \to T(\varphi)$.

We now consider several examples.

1. Let $f(t)$ be an arbitrary continuous function of t. Then, since every function $\varphi(t) \in D$ vanishes outside some finite interval, the integral

$$(1) \qquad T(\varphi) = \int_{-\infty}^{\infty} f(t)\varphi(t)\, dt$$

exists for all $\varphi \in D$. The expression (1) represents a linear functional on

D (it is left to the reader to verify that Conditions 1 and 2 of Definition 1 are actually fulfilled), i.e. (1) represents a generalized function. Thus, every "ordinary" continuous function is at the same time a generalized function also.

In this connection, two distinct continuous functions will define distinct functionals on D, i.e. they will represent distinct generalized functions. Obviously, it is sufficient to show that every continuous function $f(t)$ which is not identically zero defines a linear functional which is not equal to zero, i.e. to show that there is a function $\varphi(t) \in D$ for which

$$\int_{-\infty}^{\infty} f(t)\varphi(t) \, dt \neq 0.$$

Since $f(t) \not\equiv 0$, an interval (α, β) can be found on which $f(t)$ is different from zero and consequently keeps the same sign (for instance, positive). Let us consider the function

$$e^{-1/(t-\alpha)^2} \cdot e^{-1/(t-\beta)^2}.$$

This function is positive on the interval (α, β); it assumes the value zero at $t = \alpha, \beta$ and this is true of its derivatives of all orders. Consequently, the function $\varphi_{(\alpha,\beta)}(t)$ which is equal to $e^{-1/(t-\alpha)^2} \cdot e^{-1/(t-\beta)^2}$ on the interval (α, β) and is zero outside this interval, has continuous derivatives of all orders, i.e. it belongs to D. Furthermore,

$$\int_{-\infty}^{\infty} f(t)\varphi_{(\alpha,\beta)}(t) \, dt = \int_{\alpha}^{\beta} f(t)\varphi_{(\alpha,\beta)}(t) \, dt > 0,$$

since the function under the integral sign is positive.

2. We now set

(2) $T(\varphi) = \varphi(0).$

Formula (2) defines a linear functional on D, i.e. a generalized function. This is none other than the δ-function which we have already mentioned above. (In §23 we considered the δ-function as a functional defined on the space of *all* continuous functions. The advantages of the point of view adopted here will be manifest in the sequel.)

3. Set

$$T(\varphi) = -\varphi'(0).$$

This generalized function is called the derivative of the δ-function and is denoted by δ'. (A more general definition of the derivative of a generalized function will be given below.)

It is sometimes convenient to denote the generalized function, as in the case of an ordinary function, by the symbol $f(t)$ and say that the linear functional T is "defined" by some generalized function $f(t)$. The value $T(\varphi)$ of a linear functional T for each function $\varphi \in D$ in this context is more conveniently written in the form of a "scalar product" (f, φ). If $f(t)$ is a "proper" function,

$$(f, \varphi) = \int_{-\infty}^{\infty} f(t)\varphi(t)\, dt.$$

Frequently, especially in physics books, such an integral notation is used also for generalized functions, for example, we find:

$$\int_{-\infty}^{\infty} \delta(t)\varphi(t)\, dt = \varphi(0);$$

however, it is necessary to remember here that such an integral is only a symbolic representation of the value of the corresponding functional and that it has no other meaning.

We shall now introduce the concept of the limit of a sequence of generallized functions.

DEFINITION 2. The sequence $\{T_n\}$ is said to converge to the generalized function T if $T_n(\varphi) \to T(\varphi)$ uniformly on every "bounded" set in D. In this connection, the set $\{\varphi\} \subset D$ is said to be bounded if: 1) there exists a finite interval outside of which all the φ of this set equal zero and 2) it is possible to choose constants M_0, M_1, M_2, $\cdots$ such that

$$|\varphi| < M_0;\qquad |\varphi'| < M_1;\qquad |\varphi''| < M_2;\ \cdots$$

for all $\varphi \in \{\varphi\}$.

It is easily verified that in the sense of this definition the δ-function is the limit, for example, of such a sequence of "proper" functions:

$$f_n(t) = \begin{cases} n/2 \text{ for } -1/n < t < 1/n, \\ 0 \text{ for all other values of } t. \end{cases}$$

In exactly the same way it is easy to construct a sequence of "proper" functions converging in the indicated sense to the generalized function δ'. In general, it can be proved that every generalized function is the limit of some sequence of linear functionals defined by "proper" functions.

We shall now formulate the definition of the derivative of a generalized function.

We consider first of all the linear functional T defined by means of a differentiable function:

$$T(\varphi) = \int_{-\infty}^{\infty} f(t)\varphi(t)\, dt.$$

It is natural to call the functional defined by means of the formula

$$T'(\varphi) = \frac{dT}{dt}(\varphi) = \int_{-\infty}^{\infty} f'(t)\varphi(t)\, dt$$

the derivative T' of T.

Making use of the formula for integration by parts and taking into consideration the fact that every function $\varphi(t)$ is differentiable and equal to zero in the exterior of some finite interval, we can rewrite the last expression in the form

$$T'(\varphi) = -\int_{-\infty}^{\infty} f(t)\varphi'(t) \ dt$$

in which the derivative $f'(t)$ is no longer present in any way.

This discussion brings us to the following definition.

DEFINITION 3. Let T be a generalized function. Its derivative T' is a functional defined by the formula

$$T'(\varphi) = -T(\varphi').$$

It is clear that T' has meaning for all $\varphi \in D$ and represents a generalized function.

Derivatives of the second, third and higher orders are defined analogously.

The validity of the following assertions follows directly from the definition.

THEOREM 1. *Every generalized function has derivatives of all orders.*

THEOREM 2. *If the sequence $\{T_n\}$ converges to the generalized function T (in the sense of Definition 2), then the sequence of derivatives $\{T_n'\}$ converges to T'. In other words: every convergent series of generalized functions can be differentiated term by term any number of times.*

It is now clear that the functional δ' (see Example 3) which we called the derivative of the δ-function is in reality the derivative of $\delta(t)$ in the sense of Definition 3.

REMARK. The concept of the derivative of a generalized function can be introduced in a somewhat different way. Namely, we define the translation T_h of the functional T (where h is an arbitrary number) by setting

$$T_h(\varphi(t)) = T(\varphi(t - h))$$

(for example, $\delta_h(\varphi) = \varphi(-h)$, $\delta_h'(\varphi) = -\varphi'(-h)$, and so forth). Then it is not difficult to verify that the limit

$$\lim_{h\to 0} [(T_h - T)/h]$$

exists and is equal to the derivative T' of the functional T.

4. If the functional $T(\varphi)$ is defined by means of a differentiable function $f(t)$, then the functional T' is defined by the function $f'(t)$, i.e. the generalized derivative of a function coincides with its ordinary derivative if the latter exists.

5. Let T be defined by the function

$$f(t) = \begin{cases} 1 \text{ for } t \geq 0, \\ 0 \text{ for } t < 0. \end{cases}$$

Then

$$T(\varphi) = \int_{-\infty}^{\infty} f(t)\varphi(t)\,dt = \int_{0}^{\infty} \varphi(t)\,dt$$

and

$$T'(\varphi) = -\int_{0}^{\infty} \varphi'(t)\,dt = \varphi(0),$$

i.e. T' is the δ-function.

6. If $f(t)$ is a piecewise continuous function having a derivative at all points of continuity, its generalized derivative at points of continuity coincides with its ordinary derivative and at each point of discontinuity $t = t_0$ it is equal to the δ-function $\delta(t - t_0)$ multiplied by the magnitude of the jump of the function at this point.

7. If

$$(3) \qquad f(t) = \sum_{n=1}^{\infty} (\sin nt)/n,$$

then the generalized derivative of $f(t)$ equals

$$\sum_{n=1}^{\infty} \cos nt = -\tfrac{1}{2} + \pi \sum_{k=-\infty}^{\infty} \delta(t - 2k\pi).$$

We emphasize that the series on the left in the last equation converges in the sense of convergence for generalized functions (Definition 2). Series (3) can be differentiated term by term any number of times.

The last example shows that the concept of generalized function allows us to assign a perfectly definite meaning to the sum of a series which diverges in the ordinary sense. The same remark applies also to many divergent integrals. This situation is frequently encountered in theoretical physics where in each individual case special methods are used to give a definite meaning to a divergent series or to a divergent integral.

In an analogous way, we can introduce generalized functions of several independent variables. To do this, it is merely necessary to take as the initial space D the space of all functions of n variables which are differentiable an infinite number of times and each of which vanishes outside some sphere. All the concepts introduced above carry over automatically to this case. It is easy to verify that every generalized function of several independent variables has partial derivatives of all orders and that the result of differentiating with respect to several variables does not depend on the order of differentiation.

Chapter IV

LINEAR OPERATOR EQUATIONS

§30. Spectrum of an operator. Resolvents

Throughout this entire chapter we shall consider bounded linear operators which map a (generally speaking complex) Banach space E into itself. An operator adjoint to an operator of the type indicated will, obviously, map the conjugate space $\bar{E}$ into itself.

In the study of linear operators in finite-dimensional space an important role is played by the concepts of characteristic vector and characteristic value. In the case of n-dimensional space the concept of the characteristic value of an operator can be introduced by means of the following equivalent methods:

1) The number λ is said to be a characteristic value of the operator A if there exists a nonzero vector x (characteristic vector) such that $Ax = \lambda x$.

2) The number λ is said to be a characteristic value of the operator A if it is a root of the characteristic equation

$$Det \,|\, A - \lambda I \,| \; = 0.$$

The first of these definitions carries over without any changes to the case of infinite-dimensional space, i.e. it can be considered as the definition of characteristic values of an operator in infinite-dimensional space.

The second of the definitions formulated above does not carry over directly to the infinite-dimensional case since the concept of determinant does not have meaning in infinite-dimensional space. However, this second definition can be changed in form in the following manner. The fact that the determinant of the matrix $A - \lambda I$ is zero is equivalent to the fact that this matrix has no inverse, i.e. we can say that the characteristic values λ are those numbers for which the inverse of the operator $A - \lambda I$ does not exist. Taking this as our point of departure we introduce the following definition.

The totality of those values λ for which the inverse of the operator $A - \lambda I$ does not exist is called the *spectrum* of the operator A. The values of λ for which the operator $A - \lambda I$ has an inverse are said to be *regular*; thus, the spectrum consists of all nonregular points. In this connection, the operator $R_\lambda = (A - \lambda I)^{-1}$ itself is called the *resolvent* of the operator A. In n-dimensional space the concepts of characteristic value and nonregular, or singular, point coincide. In the general case, as the example to be introduced below will show, this is no longer the situation. The spectrum necessarily contains all the characteristic values but it may contain, besides these, other numbers also. The set of characteristic values is called the *point spectrum* of the operator A. The remaining part of the spectrum is called the *continuous spectrum*. (This is not the usual terminology. The remaining part of the spectrum consists of two parts: 1) Those values of λ for which $A - \lambda I$ has an unbounded inverse with domain dense in E form the *continuous* spectrum. 2) The remainder of the spectrum, consisting of those values of λ for which $A - \lambda I$ has an inverse whose domain is not dense in E, is called the *residual* spectrum.—Trans.)

If the point λ is regular, i.e. if the inverse of the operator $A - \lambda I$ exists, then for sufficiently small δ the operator $A - (\lambda + \delta)I$ also has an inverse (Theorem 7, §29), i.e. the point $\lambda + \delta$ is regular also. Thus, the regular points form an open set. Consequently the spectrum, which is its complement, is a closed set.

THEOREM 1. *The inverse of the operator $A - \lambda I$ exists for arbitrary λ for which $|\lambda| > \|A\|$.*

Proof. Since we obviously have

$$A - \lambda I = -\lambda \left(I - \frac{1}{\lambda} A \right),$$

then

$$R_\lambda = (A - \lambda I)^{-1} = -(1/\lambda)(I - A/2)^{-1} = -(1/\lambda) \sum_{k=0}^{\infty} A^k / \lambda^k.$$

This series converges for $\|A/\lambda\| < 1$ (see Theorem 8, §29), i.e. the operator $A - \lambda I$ has an inverse. In other words, the spectrum of the operator A is contained in a circle of radius $\|A\|$ with center at zero.

EXAMPLE. Let us consider in the space C the operator A defined by the formula

$$A(x(t)) = \mu(t)x(t),$$

where $\mu(t)$ is a fixed continuous function. We have

$$(A - \lambda I)x(t) = (\mu(t) - \lambda)x(t),$$

whence

$$(A - \lambda I)^{-1}x(t) = \{1/[\mu(t) - \lambda]\}x(t).$$

The spectrum of the operator A under consideration consists of all λ for which $\mu(t) - \lambda$ is zero for some t lying between 0 and 1, i.e. the spectrum is the totality of all values of the function $\mu(t)$. For example, if $\mu(t) = t$, then the spectrum is the closed interval [0, 1]; in this case characteristic values are absent, i.e. the operator defined as multiplication by t is an example of an operator with a purely continuous spectrum.

(1) Every linear operator, defined in a Banach space which has at least

one nonzero element, has a nonvoid spectrum. Operators exist for which the spectrum consists of a single point (for example, the operator defined as multiplication by a scalar).

(2) Theorem 1 can be sharpened in the following way. Let $r = \inf \| A^n \|^{1/n}$; then the spectrum of the operator A lies entirely in a circle with radius r and center at zero.

(3) The resolvent operators R_μ and R_λ corresponding to the points μ and λ are commutative and satisfy the relation

$$R_\mu - R_\lambda = (\mu - \lambda) R_\mu R_\lambda$$

which is easily verified by multiplying both sides of this equality by $(A - \lambda I)(A - \mu I)$. From this it follows that the derivative of R_λ with respect to λ, i.e.

$$\lim_{\Delta\lambda \to 0} (R_{\lambda+\Delta\lambda} - R_\lambda)/\Delta\lambda,$$

exists and is equal to R_λ^2.

§31. Completely continuous operators

Definition. An operator A which maps a Banach space E into itself is said to be *completely continuous* if it maps an arbitrary bounded set into a compact set.

In finite-dimensional space every bounded linear operator is completely continuous since it maps an arbitrary bounded set into a bounded set and in finite-dimensional space every bounded set is compact.

But in the case of infinite-dimensional space there always exist operators which are bounded (i.e. continuous) but which are not completely continuous; such, for example, is the identity operator in infinite-dimensional space.

An extensive class of completely continuous operators $y = Ax$ in the space $C[a, b]$ can be written in the form

$$(1) \qquad y(s) = \int_a^b K(s, t) x(t)\, dt.$$

Theorem 1. *Formula (1) defines a completely continuous operator in the space $C[a, b]$ if the function $K(s, t)$ is bounded on the square $a \leq s \leq b$, $a \leq t \leq b$ and all points of discontinuity of the function $K(s, t)$ lie on a finite number of curves*

$$t = \varphi_k(s), \qquad\qquad k = 1, 2, \cdots, n,$$

where the φ_k are continuous functions.

Remark. The distribution of points of discontinuity on the straight lines $s = constant$ is essential here. For example, for $a = 0$, $b = 1$, the kernel

$$K(s, t) = \begin{cases} 1 \text{ for } s < \frac{1}{2}, \\ 0 \text{ for } s \geq \frac{1}{2} \end{cases}$$

has for points of discontinuity all the points of the square with $s = \frac{1}{2}$. In this case transformation (1) maps the function $x(t) = 1$ into a discontinuous function.

Proof of Theorem 1. We note that under the conditions of the theorem, integral (1) exists for arbitrary s in the closed interval $a \leq s \leq b$. Let $M = \sup | K(s, t) |$ on the square $a \leq s \leq b$, $a \leq t \leq b$. Denote by G the set of points (s, t) for which $| t - \varphi_k(s) | < \epsilon/12Mn$ holds for arbitrary $k = 1, 2, \cdots, n$; denote by F the complement of G with respect to the square $a \leq s \leq b$, $a \leq t \leq b$. Since F is closed and $K(s, t)$ is continuous on F, there exists a δ such that for points (s', t), (s'', t) in F for which $| s' - s'' | < \delta$ the inequality

$$| K(s', t) - K(s'', t) | < \epsilon/3(b - a)$$

holds.

Now let s' and s'' be such that $| s' - s'' | < \delta$. Then

$$| y(s') - y(s'') | \leq \int_a^b | K(s', t) - K(s'', t) | | x(t) | dt$$

can be evaluated by integrating over the sum of the intervals

$$| t - \varphi_k(s') | < \epsilon/12Mn,$$
$$| t - \varphi_k(s'') | < \epsilon/12Mn$$

(denote this sum by A) and over the complement B of A with respect to the closed interval $[a, b]$. The length of A does not exceed $\epsilon/3M$. Therefore

$$\int_A | K(s', t) - K(s'', t) | | x(t) | dt \leq (2\epsilon/3) \| x \|.$$

The integral over B, obviously, may be estimated by

$$\int_B | K(s', t) - K(s'', t) | | x(t) | dt \leq (\epsilon/3) \| x \|.$$

Therefore,

$$| y(s') - y(s'') | \leq \epsilon \| x \|.$$

We have proved the continuity of $y(s)$ and the equicontinuity of the functions $y(s)$ corresponding to the functions $x(t)$ with bounded norm. The uniform boundedness of $y(s)$ corresponding to the $x(t)$ with bounded norm follows from the inequalities

$$| y(s) | \leq \int_a^b | K(s, t) | \, | x(t) | \, dt \leq M(b - a) \, \| x \| .$$

The Volterra Operator. If we assume that $K(s, t) = 0$ for $t > s$, then the operator (1) takes on the form

$$(2) \qquad\qquad y(s) = \int_0^s K(s, t) x(t) \, dt.$$

We shall assume the function $K(s, t)$ to be continuous for $t \leq s$. The validity of the hypothesis of Theorem 1 is easily verified for such a kernel. Consequently this operator is completely continuous.

We shall now establish several properties of completely continuous operators.

Theorem 2. *If $\{A_n\}$ is a sequence of completely continuous operators on a Banach space E which converges in norm to an operator A, then the operator A is completely continuous.*

Proof. Let us consider an arbitrary bounded sequence of elements in E: $x_1, x_2, \cdots, x_n, \cdots$; $\| x_n \| < c$. It is necessary to prove that the sequence $\{Ax_n\}$ contains a convergent subsequence.

The operator A_1 is completely continuous and therefore we can select a convergent subsequence from the sequence $\{A_1 x_n\}$. Let

$$(3) \qquad\qquad x_1^{(1)}, x_2^{(1)}, \cdots, x_n^{(1)}, \cdots$$

be the inverse images of the members of this convergent subsequence. We apply the operator A_2 to each member of the subsequence (3). Since A_2 is completely continuous, we can again select such a convergent subsequence from the sequence $\{A_2 x_n^{(1)}\}$, the inverse images of whose terms are, say,

$$x_1^{(2)}, x_2^{(2)}, \cdots, x_n^{(2)}, \cdots .$$

We apply the operator A_3 to this sequence $\{x_n^{(2)}\}$ and then in an analogous manner select a subsequence $x_1^{(3)}, x_2^{(3)}, \cdots, x_n^{(3)}, \cdots$, and so on.

We now form the diagonal sequence

$$x_1^{(1)}, x_2^{(2)}, \cdots, x_n^{(n)}, \cdots .$$

Each of the operators $A_1, A_2, \cdots, A_n, \cdots$ transforms this sequence into a convergent sequence. If we show that the operator A also transforms this sequence into a convergent sequence, then this will also establish the complete continuity of the operator A. Let us evaluate the norm of the difference $Ax_n^{(n)} - Ax_m^{(m)}$:

$$\begin{aligned}
\| Ax_n^{(n)} - Ax_m^{(m)} \| &\leq \| Ax_n^{(n)} - A_k x_n^{(n)} \| \\
&+ \| A_k x_n^{(n)} - A_k x_m^{(m)} \| + \| A_k x_m^{(m)} - A x_m^{(m)} \| \\
&\leq \| A - A_k \| (\| x_n^{(n)} \| + \| x_m^{(m)} \|) + \| A_k x_n^{(n)} - A_k x_m^{(m)} \| .
\end{aligned}$$

We choose k so that $\| A - A_k \| < \epsilon/(2c)$ and N such that the relation

$$\| A_k x_n^{(n)} - A_k x_m^{(m)} \| < \epsilon/2$$

holds for arbitrary $n > N$ and $m > N$. (This is possible because the sequence $\{A_k x_n^{(n)}\}$ converges.) Then $\| A x_n^{(n)} - A x_m^{(m)} \| < \epsilon$, i.e. the sequence $\| A x_n^{(n)} \|$ is fundamental.

Since the space E is complete, this sequence converges; this proves the theorem.

THEOREM 3. *The adjoint operator of a completely continuous operator is completely continuous.*

Proof. Let the operator A map the Banach space E into itself. Then the operator A^*, adjoint to A, maps the space $\bar{E}$ into itself. Inasmuch as an arbitrary bounded set can be enclosed in a sphere, the theorem will be proved if we prove that the image of the sphere $\bar{S}$ belonging to the space $\bar{E}$ (i.e. the set $A^*\bar{S}$) is compact. It is obviously sufficient to carry out the proof for the case of the unit sphere.

We introduce into the space $\bar{E}$ the auxiliary metric obtained by setting for any two functionals g', g'' in $\bar{E}$,

$$(4) \qquad \rho(g', g'') = \sup \{| g'(y) - g''(y) |; y \in AS\}.$$

Since the operator A is completely continuous by hypothesis, the image AS of the unit sphere S is compact in the space E. If $y \in AS$, then $\| y \| \leq \| A \|$. Consequently if $g \in \bar{S}$, then $| g(y) | \leq \| g \| \| y \| \leq \| A \|$, i.e. the functionals $g(y) \in \bar{S}$ are uniformly bounded in the sense of metric (4).

Further, we have

$$| g(y') - g(y'') | \leq \| g \| \| y' - y'' \| \leq \| y' - y'' \|,$$

i.e. the functionals $g(y) \in \bar{S}$ are equicontinuous. In accordance with Arzelà's theorem, §17, this means that the set $\bar{S}$ is compact in the sense of metric (4). But since

$$\| A^*g' - A^*g'' \| \leq \sup \{| g'(Ax) - g''(Ax) |; x \in S\},$$

the compactness of the set $\bar{S}$ in the sense of metric (4) implies the compactness of the set $A^*\bar{S}$ in the sense of the original metric of the space $\bar{E}$.

THEOREM 4. *If A is a completely continuous operator and B is a bounded operator, then the operators AB and BA are also completely continuous.*

Proof. If M is a bounded set in E, then the set BM is also bounded. Consequently, the set ABM is compact and this means that the operator AB is completely continuous. Further, if M is bounded, then AM is compact; but then in virtue of the continuity of B, the set BAM is also compact. This completes the proof of the theorem.

COROLLARY. A completely continuous operator A cannot have a bounded inverse in an infinite-dimensional space E.

In fact, in the contrary case the identity operator $I = AA^{-1}$ would be completely continuous in E, which is impossible.

We shall now study the spectrum of a completely continuous operator.

THEOREM 5. *Every completely continuous operator A in the Banach space E has for arbitrary $\rho > 0$ only a finite number of linearly independent characteristic vectors which correspond to the characteristic values whose absolute values are greater than ρ.*

Proof. Let us assume that this is not the situation, i.e. let us assume that there exists an infinite number of linearly independent characteristic vectors x_n ($n = 1, 2, \cdots$), satisfying the condition

$$(5) \qquad Ax_n = \lambda_n x_n, \qquad |\lambda_n| > \rho > 0, \qquad (n = 1, 2, \cdots).$$

Consider in E the subspace E_1 generated by the vectors x_n. In this subspace the operator A has a bounded inverse. In fact, vectors of the form

$$(6) \qquad x = \alpha_1 x_1 + \alpha_2 x_2 + \cdots + \alpha_k x_k$$

are everywhere dense in E_1. For these vectors,

$$Ax = \alpha_1 \lambda_1 x_1 + \alpha_2 \lambda_2 x_2 + \cdots + \alpha_k \lambda_k x_k$$

and consequently,

$$(7) \qquad A^{-1}x = (\alpha_1/\lambda_1)x_1 + (\alpha_2/\lambda_2)x_2 + \cdots + (\alpha_k/\lambda_k)x_k$$

and

$$\| A^{-1}x \| \leq (1/\rho) \| x \|.$$

Consequently the operator A^{-1}, defined for vectors of the form (6) by means of formula (7), can be extended by continuity (and consequently with preservation of norm) to all of E_1. The operator A, being completely continuous in E, is completely continuous in E_1 also. But according to the corollary to the preceding theorem a completely continuous operator cannot have a bounded inverse in infinite-dimensional space. The contradiction thus obtained proves the theorem. It follows from this theorem that every nonzero characteristic value of a completely continuous operator has only finite multiplicity and these characteristic values form a bounded set which cannot have a single limit point distinct from the origin of coordinates. We have thus obtained a characterization of the point spectrum of a completely continuous operator. It will follow from the results of the next section that a completely continuous operator cannot have a continuous spectrum.

§32. Linear operator equations. The Fredholm theorems

In this section we shall consider equations of the form

$$(1) \qquad y = x - Ax,$$

where A is a completely continuous operator which maps a Banach space E into itself. We shall show that for such equations a number of theorems are valid which are analogous to the known theorems for systems of linear algebraic equations. First of all the corresponding theory was developed by Fredholm for integral equations of the form

$$(2) \qquad \varphi(t) = f(t) + \int_a^b K(s, t)\varphi(s) \, ds,$$

where $f(t)$ and $K(s, t)$ are given continuous functions.

It was established in the preceding section that the Fredholm operator

$$y(t) = \int_a^b K(s, t)x(s) \, ds$$

is completely continuous; thus, equation (1) actually is a generalization of the Fredholm integral equation (2).

THEOREM 1. *Let A be a completely continuous operator which maps the Banach space E into itself. If the equation $y = x - Ax$ is solvable for arbitrary y, then the equation $x - Ax = 0$ has no solutions distinct from the zero solution.*

Proof. Let us assume the contrary: suppose there exists an element $x_1 \neq 0$ such that $x_1 - Ax_1 = Tx_1 = 0$. Those elements x for which $Tx = 0$ form a linear subspace E_1 of the space E. Let us denote by E_n the subspace consisting of the elements which satisfy the condition $T^n x = 0$.

It is clear that the subspaces E_n form a nondecreasing subsequence

$$E_1 \subseteq E_2 \subseteq \cdots \subseteq E_n \subseteq \cdots .$$

We shall show that the equality sign cannot hold at any point of this chain of inclusion relations. In fact, since by assumption $x_1 \neq 0$ and the equation $y = Tx$ is solvable for arbitrary y, we can find a sequence of elements distinct from zero $x_1, x_2, \cdots, x_n, \cdots$, such that

$$Tx_2 = x_1$$
$$Tx_3 = x_2$$
$$\cdots\cdots\cdots$$
$$Tx_n = x_{n-1}$$
$$\cdots\cdots\cdots .$$

The element x_n belongs to the subspace E_n for each n but it does not belong to the subspace E_{n-1}. In fact,

$$T^n x_n = T^{n-1} x_{n-1} = \cdots = Tx_1 = 0,$$

but

$$T^{n-1} x_n = T^{n-2} x_{n-1} = \cdots = Tx_2 = x_1 \neq 0.$$

All the subspaces E_n are linear and closed; therefore for arbitrary n there exists an element $y_{n+1} \in E_{n+1}$ such that

$$\| y_{n+1} \| = 1 \quad \text{and} \quad \rho(y_{n+1}, E_n) \geq \tfrac{1}{2},$$

where $\rho(y_{n+1}, E_n)$ denotes the distance from y_{n+1} to the space E_n, i.e.

$$\rho(y_{n+1}, E_n) = \inf \{\| y_{n+1} - x \|; x \in E_n\}.$$

[In fact, if we let $\rho(x_{n+1}, E_n) = \alpha$, then we can find an element $\tilde{x} \in E_n$ such that $\| x_{n+1} - \tilde{x} \| < 2\alpha$; at the same time we clearly have $\rho(x_{n+1} - \tilde{x}, E_n) = \rho(x_{n+1}, E_n) = \alpha$. We can then set $y_{n+1} = (x_{n+1} - \tilde{x})/\| x_{n+1} - \tilde{x} \|$.]
Consider the sequence $\{Ay_k\}$. We have (assuming $p > q$):

$$\| Ay_p - Ay_q \| = \| y_p - (y_q + Ty_p - Ty_q) \| \geq \tfrac{1}{2},$$

since $y_q + Ty_p - Ty_q \in E_{p-1}$. It is clear from this that the sequence $\{Ay_k\}$ cannot contain any convergent subsequence, which contradicts the complete continuity of the operator A. The contradiction thus obtained proves the theorem.

COROLLARY 1. If the equation $y = x - Ax$ is solvable for arbitrary y, then it has a unique solution for each y, i.e. the operator $I - A$ has an inverse in this case.

In fact, if the equation $y = x - Ax$ had two distinct solutions for some y, say x_1 and x_2, then the equation $x - Ax = 0$ would have a nonzero solution $x_1 - x_2$, which contradicts Theorem 1.

In the sequel it is convenient to consider together with the equation $y = x - Ax$ the equation $h = f - A^*f$ which is adjoint to it, where A^* is the adjoint of A and h, f are elements of the Banach space $\bar{E}$, the conjugate of E.

For the adjoint equation we can formulate the following result.

COROLLARY 2. If the equation $h = f - A^*f$ is solvable for all h, then the equation $f - A^*f = 0$ has only the zero solution.

This assertion is obtained from Theorem 1 if we recall that the operator adjoint to a completely continuous operator is completely continuous (Theorem 2, §31) and a space conjugate to a Banach space is itself a Banach space.

THEOREM 2. *A necessary and sufficient condition that the equation* $y = x - Ax$ *be solvable is that the following condition be fulfilled:* $f(y) = 0$ *for all f for which* $f - A^*f = 0$.

Proof. 1. If we assume the equation $y = x - Ax$ is solvable, then $f(y) = f(x) - f(Ax) = f(x) - A^*f(x)$, i.e. $f(y) = 0$ for all f which satisfy the condition $f - A^*f = 0$.

2. Now let $f(y)$ equal zero for all f which satisfy the equation $f - A^*f = 0$. For each of these functionals f we shall consider the set L_f of elements for

which f takes on the value zero. Then our assertion is equivalent to the fact that the set $\cap L_f$ consists only of the elements of the form $x - Ax$, i.e. it is necessary for us to prove that an element y_1 which cannot be represented in the form $x - Ax$ cannot be contained in $\cap L_f$. To do this we shall show that for such an element y_1 we can construct a functional f_1 which satisfies the conditions

$$f_1(y_1) \neq 0, \qquad f_1 - A^*f_1 = 0.$$

These conditions are equivalent to the following:

$$f_1(y_1) \neq 0, \qquad f_1(x - Ax) = 0 \quad \text{for all } x.$$

In fact,

$$(f_1 - A^*f_1)(x) = f_1(x) - A^*f_1(x) = f_1(x) - f_1(Ax) = f_1(x - Ax).$$

Let G_0 be the subspace consisting of all elements of the form $x - Ax$. Consider the subspace $\{G_0, y_1\}$, i.e. the set of elements of the form $z + \alpha y_1$, $z \in G_0$, and define the linear functional f_1 on this subspace by setting

$$f_1(z + \alpha y_1) = \alpha.$$

Extending this functional to the whole space E (which is possible by virtue of the Hahn-Banach theorem) we do obtain a linear functional satisfying the required conditions. This completes the proof of the theorem.

COROLLARY. If the equation $f - A^*f = 0$ does not have nonzero solutions, then the equation $x - Ax = y$ is solvable for all y.

We shall now establish the analogue of Theorem 2 for the adjoint equation.

THEOREM 3. *A necessary and sufficient condition that the equation*

(3) $$h = f - A^*f$$

be solvable is that $h(x) = 0$ *for all x for which* $x - Ax = 0$.

Proof. 1. If the equation $h = f - A^*f$ is solvable, then

$$h(x) = f(x) - A^*f(x) = f(x - Ax),$$

i.e. $h(x) = 0$ if $x - Ax = 0$.

2. Now let $h(x) = 0$ for all x which satisfy the equation $x - Ax = 0$. We shall show that equation (3) is solvable. We shall construct the functional f on the set F of all elements of the form $y = x - Ax$ by setting

(4) $$f(Tx) = h(x).$$

This equation in fact defines a linear functional. First of all we note that the value of the functional f is defined uniquely for each y because if $Tx_1 = Tx_2$, then $h(x_1) = h(x_2)$. It is easy to verify the linearity of the functional (4). This functional can be extended to the whole space E. We obtain

$$f(Tx) = T^*f(x) = h(x),$$

i.e. this functional is a solution of equation (3).

COROLLARY. If the equation $x - Ax = 0$ has no nonzero solutions, then the equation $f - A^*f = h$ is solvable for all h.

The following theorem is the converse of Theorem 1.

THEOREM 4. *If the equation $x - Ax = 0$ has $x = 0$ for its only solution, then the equation $x - Ax = y$ has a solution for all y.*

Proof. If the equation $x - Ax = 0$ has only one solution, then by virtue of the corollary to the preceding theorem the equation $f - A^*f = h$ is solvable for all h. Then, by Theorem 1, Corollary 2, the equation $f - A^*f = 0$ has only the zero solution. Hence the Corollary to Theorem 2 implies the solvability of the equation $y = x - Ax$ for arbitrary y.

Theorems 1 and 4 show that for the equation

$$(1) \qquad y = x - Ax$$

only the following two cases are possible:

1) Equation (1) has a unique solution for each y, i.e. the operator $I - A$ has an inverse.

2) The equation $x - Ax = 0$ has a nonzero solution, i.e. the number $\lambda = 1$ is a characteristic value for the operator A.

It is clear that all the results obtained above for the equation $y = x - Ax$ carry over automatically also to the equation $y = x - \lambda Ax$, where λ is an arbitrary number. (For such an equation the adjoint equation will be $h = f - \bar{\lambda}A^*f$.) It follows that either the operator $I - \lambda A$ has an inverse or that the number $1/\lambda$ is a characteristic value of the operator A. In other words, in the case of a completely continuous operator, an arbitrary number is either a regular point or a characteristic value, i.e. a completely continuous operator has only a point spectrum. The point spectrum of a completely continuous operator was studied in the preceding section (Theorem 5).

THEOREM 5. *The dimension n of the space N whose elements are the solutions of the equation $x - Ax = 0$ is equal to the dimension of the subspace N^* whose elements are the solutions of the equation $f - A^*f = 0$.*

Proof. We note first that in virtue of the complete continuity of the operators A and A^*, the subspaces N and N^* are finite-dimensional. Let $\{x_1, x_2, \cdots, x_n\}$ be a basis for the subspace N and let $\{f_1, f_2, \cdots, f_\nu\}$ be a basis for the subspace N^*. According to the Hahn-Banach theorem on the extension of a linear functional, n functionals $\varphi_1, \varphi_2, \cdots, \varphi_n$ can be constructed which satisfy the conditions

$$\varphi_j(x_i) = \begin{cases} 1 \text{ for } i = j, \\ 0 \text{ for } i \neq j. \end{cases}$$

We now consider the functionals $f_1, f_2, \cdots, f_\nu$ and construct for each of them the corresponding subspace L_i, defined by the condition $f_i(x) = 0$. Since the functionals f_i are linearly independent, the corresponding sets L_i are distinct; and since each of the subspaces L_i has index 1, a relation of the form $L_j \subset L_i$ is likewise impossible.

Let us consider the sets

$$L_1 \cap L_2 \cap \cdots \cap L_{\nu-1} \setminus L_1 \cap L_2 \cap \cdots \cap L_\nu = M_\nu \neq 0,$$

$$\cdots\cdots\cdots\cdots\cdots\cdots\cdots\cdots\cdots\cdots\cdots\cdots$$

$$L_1 \cap \cdots \cap L_{i-1} \cap L_{i+1} \cap \cdots \cap L_\nu \setminus L_1 \cap L_2 \cap \cdots \cap L_\nu = M_i \neq 0,$$

$$\cdots\cdots\cdots\cdots\cdots\cdots\cdots\cdots\cdots\cdots\cdots\cdots$$

$$L_2 \cap L_3 \cap \cdots \cap L_\nu \setminus L_1 \cap L_2 \cap \cdots \cap L_\nu = M_1 \neq 0.$$

The set M_i consists of those elements which belong to all L_j, except L_i. It is clear that the sets M_i are disjoint. Since all the functionals $f_1, f_2, \cdots, f_\nu$, except f_i, assume the value zero on M_i and since M_i is a linear manifold, it is possible to choose a point z_i in each M_i to satisfy the following conditions:

$$(5) \qquad f_j(z_i) = \begin{cases} 1 \text{ for } i = j, \\ 0 \text{ for } i \neq j. \end{cases}$$

After these auxiliary constructions we pass directly to the proof of the assertion of the theorem.

We assume first that $\nu > n$.

We construct the operator

$$R(x) = Ax + \sum_{j=1}^n \varphi_j(x) z_j.$$

This operator is completely continuous since it is the sum of a finite number of completely continuous operators. Further, let us consider the operator

$$W(x) = x - R(x) = x - Ax - \sum_{j=1}^n \varphi_j(x) z_j.$$

We shall show that if $W(x_0) = 0$, then $x_0 = 0$.

Let us apply the functional f_i to $W(x_0)$:

$$(6) \qquad f_i(W(x_0)) = f_i(Tx_0) - \sum_{j=1}^n \varphi_j(x_0) f_i(z_j) = f_i(Tx_0) - \varphi_i(x_0).$$

Since, by hypothesis, f_i satisfies the equation $T^* f_i = 0$, $f_i(Tx) = 0$ for all x ($i = 1, 2, \cdots, \nu$). Besides, $f_i(W(x_0)) = 0$ since, by assumption, $Wx_0 = 0$. Therefore from (6) it follows that

$$(7) \qquad\qquad \varphi_i(x_0) = 0 \qquad\qquad (i = 1, 2, \cdots, n).$$

But then the equation $Wx_0 = 0$ implies that $Tx_0 = 0$, i.e. $x_0 \in N$; since the elements $x_1, x_2, \cdots, x_n$ form a basis in N, we have

$$x_0 = \alpha_1 x_1 + \alpha_2 x_2 + \cdots + \alpha_n x_n.$$

If we apply the functional φ_i to this equation and make use of equations (7), we see that $\alpha_i = \varphi_i(x_0)$. This and (7) imply that $x_0 = 0$. We have shown that the equation $Wx = 0$ has only the zero solution. From this it follows (Theorem 4) that the equation $Wx = y$ is solvable for all y and, in particular, for $y = z_{n+1}$.

Let x be a solution of the equation $Wx = z_{n+1}$. Then, on the one hand, it follows from the fact that $T^*f_{n+1} = 0$ and from (5) that

$$f_{n+1}(Wx) = f_{n+1}(Tx) - f_{n+1}[\textstyle\sum_{j=1}^{n} \varphi_j(x)z_j] = 0;$$

on the other hand, according to (5), that

$$f_{n+1}(z_{n+1}) = 1.$$

The contradiction thus obtained proves that $\nu \leq n$.

Let us now assume that $n > \nu$. We construct the operator W^* by setting

$$(8) \qquad W^*f = T^*f - \textstyle\sum_{i=1}^{\nu} f(z_i)\varphi_i .$$

We shall prove that if $W^*f = 0$, then $f = 0$. We have

$$(T^*f)x_i = 0 \qquad\qquad (i = 1, 2, \cdots, n)$$

for all f since $x_i \in N$. Therefore, if $W^*f = 0$, applying both sides of (8) to x_i, we obtain:

$$(9) \qquad \textstyle\sum_{k=1}^{\nu} f(z_k)\varphi_k(x_i) = f(z_i) = 0 \qquad (i = 1, 2, \cdots, \nu).$$

Thus, the equation $W^*f = 0$ reduces to $T^*f = 0$, i.e. $f \in N^*$. Consequently, f can be written in the form

$$f = \beta_1 f_1 + \beta_2 f_2 + \cdots + \beta_\nu f_\nu .$$

If we apply both sides of this equation to the element z_i we see that $\beta_i = f(z_i)$. From this and (9) it follows that $f = 0$. This implies that the equation $W^*f = g$ is solvable for all g.

Let f be such that $W^*f = \varphi_{\nu+1}$. Then, on the one hand,

$$(W^*f)(x_{\nu+1}) = (T^*f)(x_{\nu+1}) - \textstyle\sum_{i=1}^{\nu} f(z_i)\varphi_i(x_{\nu+1}) = 0,$$

and on the other,

$$\varphi_{\nu+1}(x_{\nu+1}) = 1.$$

We have thus arrived at a contradiction, which proves that $n \leq \nu$. So finally, $n = \nu$. This completes the proof of the theorem.

LIST OF SYMBOLS

LIST OF DEFINITIONS

LIST OF THEOREMS

BASIC LITERATURE

ALEKSANDROV, P. S.
  Introduction to the General Theory of Sets and Functions, Moscow-Leningrad, Gosudarstv. Izdat. Tehn.-Teor. Lit., 1948.
BANACH, S.
  Théorie des Opérations Linéaires, Warsaw, 1932 (Monografje Matematyczne, Vol. I).
FRAENKEL, A. A.
  Abstract Set Theory, Amsterdam, North-Holland, 1953.
  Einleitung in die Mengenlehre, New York, Dover, 1946.
GRAVES, L. M.
  The Theory of Functions of Real Variables, 2nd edition, New York, McGraw-Hill, 1956.
HALL, D. W. AND SPENCER, G. L.
  Elementary Topology, New York, Wiley, 1955.

HAUSDORFF, F.
  Mengenlehre, 3rd edition, New York, Dover, 1944.
HILLE, E.
  Functional Analysis and Semi-Groups, New York, AMS Coll. Pub., Vol. XXXI, 1948.
JEFFERY, R. L.
  The Theory of Functions of a Real Variable, Toronto, University of Toronto Press, 1951.
KAMKE, E.
  Theory of Sets, New York, Dover, 1950.
KANTOROVICH, L. V.
  Functional Analysis and Applied Mathematics, Los Angeles, National Bureau of Standards, 1952.
KELLEY, J. L.
  General Topology, New York, Van Nostrand, 1955.
KURATOWSKI, K.
  Introduction to the Theory of Sets and Topology, Warsaw, 1955 (Biblioteka Matematyczna).
LOVITT, W. V.
  Linear Integral Equations, New York, Dover, 1950.
LYUSTERNIK, L. A. AND SOBOLEV, V. I.
  The Elements of Functional Analysis, Moscow-Leningrad, Gosudarstv. Izdat. Tehn.-Teor. Lit., 1951.
NATANSON, I. P.
  Theory of Functions of a Real Variable, New York, Ungar, 1955.
NEWMAN, M. H. A.
  Elements of the Topology of Plane Sets of Points, 2nd edition, Cambridge, 1951.
RIESZ, F. AND SZ.-NAGY, B.
  Functional Analysis, New York, Ungar, 1955.
SIERPIŃSKI, W.
  General Topology, Toronto, University of Toronto Press, 1952.
ŠILOV, G. E.
  Introduction to the Theory of Linear Spaces, Moscow-Leningrad, Gosudarstv. Izdat. Tehn.-Teor. Lit., 1952.
WILDER, R. L.
  Introduction to the Foundations of Mathematics, New York, Wiley, 1952.
ZAANEN, A. C.
  Linear Analysis, New York, Interscience, 1953.

INDEX

Elements of the Theory of Functions and Functional Analysis

Volume 2: Measure.
The Lebesgue Integral.
Hilbert Space

CONTENTS

CHAPTER VIII

SQUARE INTEGRABLE FUNCTIONS

CHAPTER IX

ABSTRACT HILBERT SPACE. INTEGRAL EQUATIONS WITH SYMMETRIC KERNEL

PREFACE

This book is the second volume of *Elements of the Theory of Functions and Functional Analysis* (the first volume was *Metric and Normed Spaces*, Graylock Press, 1957). Most of the second volume is devoted to an exposition of measure theory and the Lebesgue integral. These concepts, particularly the concept of measure, are discussed with some degree of generality. However, in order to achieve greater intuitive insight, we begin with the definition of plane Lebesgue measure. The reader who wishes to do so may, after reading §33, go on at once to Ch. VI and then to the Lebesgue integral, if he understands the measure relative to which this integral is taken to be the usual linear or plane Lebesgue measure.

The exposition of measure theory and the Lebesgue integral in this volume is based on the lectures given for many years by A. N. Kolmogorov in the Department of Mathematics and Mechanics at the University of Moscow. The final draft of the text of this volume was prepared for publication by S. V. Fomin.

The content of Volumes 1 and 2 is approximately that of the course Analysis III given by A. N. Komogorov for students in the Department of Mathematics.

For convenience in cross-reference, the numbering of chapters and sections in the second volume is a continuation of that in the first.

Corrections to Volume 1 have been listed in a supplement at the end of Volume 2.

A. N. KOLMOGOROV
S. V. FOMIN

January 1958

TRANSLATORS' NOTE

In order to enhance the usefulness of this book as a text, a complete set of exercises (listed at the end of each section) has been prepared by H. Kamel. It is hoped that the exercises will not only test the reader's understanding of the text, but will also introduce or extend certain topics which were either not mentioned or briefly alluded to in the original.

The material which appeared in the original in small print has been enclosed by stars (★) in this translation.

Chapter V

MEASURE THEORY

The measure $\mu(A)$ of a set A is a natural generalization of the following concepts:

1) The length $l(\Delta)$ of a segment Δ.

2) The area $S(F)$ of a plane figure F.

3) The volume $V(G)$ of a three-dimensional figure G.

4) The increment $\varphi(b) - \varphi(a)$ of a nondecreasing function $\varphi(t)$ on a half-open interval $[a, b)$.

5) The integral of a nonnegative function over a one-, two-, or three-dimensional region, etc.

The concept of the measure of a set, which originated in the theory of functions of a real variable, has subsequently found numerous applications in the theory of probability, the theory of dynamical systems, functional analysis and other branches of mathematics.

In §33 we discuss the concept of measure for plane sets, based on the area of a rectangle. The general theory of measure is taken up in §§35–39. The reader will easily notice, however, that all the arguments and results of §33 are general in character and are repeated with no essential changes in the abstract theory.

§33. The measure of plane sets

We consider the collection $\mathfrak{S}$ of sets in the plane (x, y), each of which is defined by an inequality of the form

$$a \leq x \leq b,$$
$$a < x \leq b,$$
$$a \leq x < b,$$
$$a < x < b,$$

and by an inequality of the form

$$c \leq y \leq d,$$
$$c < y \leq d,$$
$$c \leq y < d,$$
$$c < y < d,$$

1

where a, b, c and d are arbitrary real numbers. We call the sets of $\mathfrak{S}$ *rectangles*. A closed rectangle defined by the inequalities

$$a \leq x \leq b; \quad c \leq y \leq d$$

is a rectangle in the usual sense (together with its boundary) if $a < b$ and $c < d$, or a segment (if $a = b$ and $c < d$ or $a < b$ and $c = d$), or a point (if $a = b, c = d$), or, finally, the empty set (if $a > b$ or $c > d$). An open rectangle

$$a < x < b; \quad c < y < d$$

is a rectangle without its boundary or the empty set, depending on the relative magnitudes of a, b, c and d. Each of the rectangles of the remaining types (we shall call them half-open rectangles) is either a proper rectangle with one, two or three sides included, or an interval, or a half-interval, or, finally, the empty set.

The measure of a rectangle is defined by means of its area from elementary geometry as follows:

a) The measure of the empty set $\emptyset$ is zero.

b) The measure of a nonempty rectangle (closed, open or half-open), defined by the numbers a, b, c and d, is equal to

$$(b - a)(d - c).$$

Hence, we have assigned to each rectangle P a number $m(P)$—the measure of P. The following conditions are obviously satisfied:

1) The measure $m(P)$ is real-valued and nonnegative.

2) The measure $m(P)$ is additive, i.e., if $P = \bigcup_{k=1}^{n} P_k$ and $P_i \cap P_k = \emptyset$ for $i \neq k$, then

$$m(P) = \sum_{k=1}^{n} m(P_k).$$

Our problem is to extend the measure $m(P)$, defined above for rectangles, to a more general class of sets, while retaining Properties 1) and 2).

The first step consists in extending the concept of measure to the so called elementary sets. We shall call a plane set *elementary* if it can be written, in at least one way, as a union of a finite number of pairwise disjoint rectangles.

In the sequel we shall need

THEOREM 1. *The union, intersection, difference and symmetric difference of two elementary sets is an elementary set.*

Proof. It is clear that the intersection of two rectangles is again a rectangle. Therefore, if

$$A = \bigcup_k P_k, \quad B = \bigcup_j Q_j$$

are elementary sets, then

$$A \cap B = \bigcup_{k,j} (P_k \cap Q_j)$$

is also an elementary set.

It is easily verified that the difference of two rectangles is an elementary set. Consequently, subtraction of an elementary set from a rectangle yields an elementary set (as the intersection of elementary sets). Now let A and B be two elementary sets. There is clearly a rectangle P containing both sets. Then

$$A \cup B = P \setminus [(P \setminus A) \cap (P \setminus B)]$$

is an elementary set. Since

$$A \setminus B = A \cap (P \setminus B),$$
$$A \triangle B = (A \cup B) \setminus (A \cap B),$$

it follows that the difference and the symmetric difference of two elementary sets are elementary sets. This proves the theorem.

We now define the measure $m'(A)$ of an elementary set A as follows: If

$$A = \bigcup_k P_k,$$

where the P_k are pairwise disjoint rectangles, then

$$m'(A) = \sum_k m(P_k).$$

We shall prove that $m'(A)$ is independent of the way in which A is represented as a union of rectangles. Let

$$A = \bigcup_k P_k = \bigcup_j Q_j,$$

where P_k and Q_j are rectangles, and $P_i \cap P_k = \emptyset$, $Q_i \cap Q_k = \emptyset$ for $i \neq k$. Since $P_k \cap Q_j$ is a rectangle, in virtue of the additivity of the measure for rectangles we have

$$\sum_k m(P_k) = \sum_{k,j} m(P_k \cap Q_j) = \sum_j m(Q_j).$$

It is easily seen that the measure of elementary sets defined in this way is nonnegative and additive.

A property of the measure of elementary sets important for the sequel is given by

THEOREM 2. *If A is an elementary set and $\{A_n\}$ is a countable (finite or denumerable) collection of elementary sets such that*

$$A \subseteq \bigcup_n A_n,$$

then

(1) $$m'(A) \leq \sum_n m'(A_n).$$

Proof. For arbitrary $\epsilon > 0$ and given A there obviously exists a closed elementary set $\bar{A}$ contained in A and satisfying the condition

$$m'(\bar{A}) \geq m'(A) - \epsilon/2.$$

[It is sufficient to replace each of the k rectangles P_i whose union is A by a closed rectangle contained in P_i and having an area greater than $m(P_i) - \epsilon/2^{k+1}$.]

Furthermore, for each n there is an open elementary set $\tilde{A}_n$ containing A_n and such that

$$m'(\tilde{A}_n) \leq m'(A_n) + \epsilon/2^{n+1}.$$

It is clear that

$$\bar{A} \subseteq \bigcup_n \tilde{A}_n.$$

Since $\bar{A}$ is compact, by the Heine-Borel theorem (see §18, Theorem 4) $\{\tilde{A}_n\}$ contains a finite subsequence $\tilde{A}_{n_1}, \cdots, \tilde{A}_{n_s}$ which covers $\bar{A}$. Obviously,

$$m'(\bar{A}) \leq \sum_{i=1}^{s} m'(\tilde{A}_{n_i}).$$

[In the contrary case $\bar{A}$ would be covered by a finite number of rectangles the sum of whose areas is less than $m'(\bar{A})$, which is clearly impossible.] Therefore,

$$\begin{aligned}
m'(A) \leq m'(\bar{A}) + \epsilon/2 &\leq \sum_{i=1}^{s} m'(\tilde{A}_{n_i}) + \epsilon/2 \\
&\leq \sum_n m'(\tilde{A}_n) + \epsilon/2 \\
&\leq \sum_n m'(A_n) + \sum_n \epsilon/2^{n+1} + \epsilon/2 \\
&= \sum_n m'(A_n) + \epsilon.
\end{aligned}$$

Since $\epsilon > 0$ is arbitrary, (1) follows.

The class of elementary sets does not exhaust all the sets considered in elementary geometry and classical analysis. It is therefore natural to pose the question of extending the concept of measure, while retaining its fundamental properties, to a class of sets wider than the finite unions of rectangles with sides parallel to the coordinate axes.

This problem was solved, in a certain sense definitively, by Lebesgue in the early years of the twentieth century.

In presenting the Lebesgue theory of measure it will be necessary to consider not only finite, but also infinite unions of rectangles.

In order to avoid infinite values of the measure, we restrict ourselves in the sequel to sets contained in the square $E = \{0 \leq x \leq 1; 0 \leq y \leq 1\}$.

We define two functions, $\mu^*(A)$ and $\mu_*(A)$, on the class of all sets A contained in E.

DEFINITION 1. *The outer measure $\mu^*(A)$ of a set A is*

$$\mu^*(A) = \inf \{ \sum m(P_k); A \subset \cup P_k \},$$

where the lower bound is taken over all coverings of A by countable collections of rectangles.

DEFINITION 2. *The inner measure $\mu_*(A)$ of a set A is*

$$\mu_*(A) = 1 - \mu^*(E \setminus A).$$

It is easy to see that

$$\mu_*(A) \leq \mu^*(A)$$

for every set A.

For, suppose that there is a set $A \subset E$ such that

$$\mu_*(A) > \mu^*(A),$$

i.e.,

$$\mu^*(A) + \mu^*(E \setminus A) < 1.$$

Then there exist sets of rectangles $\{P_i\}$ and $\{Q_k\}$ covering A and $E \setminus A$, respectively, such that

$$\sum_i m(P_i) + \sum_k m(Q_k) < 1.$$

Denoting the union of the sets $\{P_i\}$ and $\{Q_k\}$ by $\{R_j\}$, we see that

$$E \subseteq \cup_j R_j, \qquad m(E) > \sum_j m(R_j).$$

This contradicts Theorem 2.

DEFINITION 3. *A set A is said to be measurable (in the sense of Lebesgue) if*

$$\mu_*(A) = \mu^*(A).$$

The common value $\mu(A)$ of the outer and inner measures of a measurable set A is called its Lebesgue measure.

We shall derive the fundamental properties of Lebesgue measure and measurable sets, but first we prove the following property of outer measure.

THEOREM 3. *If*

$$A \subseteq \cup_n A_n,$$

where $\{A_n\}$ is a countable collection of sets, then

$$\mu^*(A) \leq \sum_n \mu^*(A_n).$$

Proof. According to the definition of outer measure, for each n and every

$\epsilon > 0$ there is a countable collection of rectangles $\{P_{nk}\}$ such that $A_n \subseteq \bigcup_k P_{nk}$ and

$$\sum_k m(P_{nk}) \leq \mu^*(A_n) + \epsilon/2^n.$$

Then

$$A \subseteq \bigcup_n \bigcup_k P_{nk},$$

and

$$\mu^*(A) \leq \sum_n \sum_k m(P_{nk}) \leq \sum_n \mu^*(A_n) + \epsilon.$$

This completes the proof of the theorem.

Theorem 4 below shows that the measure m' introduced for elementary sets coincides with the Lebesgue measure of such sets.

THEOREM 4. *Every elementary set A is measurable, and $\mu(A) = m'(A)$.*

Proof. If A is an elementary set and $P_1, \cdots, P_k$ are rectangles whose union is A, then by definition

$$m'(A) = \sum_{i=1}^{k} m(P_i).$$

Since the rectangles P_i cover A,

$$\mu^*(A) \leq \sum_i m(P_i) = m'(A).$$

But if $\{Q_j\}$ is an arbitrary countable set of rectangles covering A, then, by Theorem 2, $m'(A) \leq \sum_j m(Q_j)$. Consequently, $m'(A) \leq \mu^*(A)$. Hence, $m'(A) = \mu^*(A)$.

Since $E \setminus A$ is also an elementary set, $m'(E \setminus A) = \mu^*(E \setminus A)$. But

$$m'(E \setminus A) = 1 - m'(A), \quad \mu^*(E \setminus A) = 1 - \mu_*(A).$$

Hence,

$$m'(A) = \mu_*(A).$$

Therefore,

$$m'(A) = \mu^*(A) = \mu_*(A) = \mu(A).$$

Theorem 4 implies that Theorem 2 is a special case of Theorem 3.

THEOREM 5. *In order that a set A be measurable it is necessary and sufficient that it have the following property: for every $\epsilon > 0$ there exists an elementary set B such that*

$$\mu^*(A \triangle B) < \epsilon.$$

In other words, the measurable sets are precisely those which can be approximated to an arbitrary degree of accuracy by elementary sets. For the proof of Theorem 5 we require the following

LEMMA. *For arbitrary sets A and B*,

$$| \mu^*(A) - \mu^*(B) | \leq \mu^*(A \Delta B).$$

Proof of the Lemma. Since

$$A \subset B \cup (A \Delta B),$$

it follows that

$$\mu^*(A) \leq \mu^*(B) + \mu^*(A \Delta B).$$

Hence the lemma follows if $\mu^*(A) \geq \mu^*(B)$. If $\mu^*(A) \leq \mu^*(B)$, the lemma follows from the inequality

$$\mu^*(B) \leq \mu^*(A) + \mu^*(A \Delta B),$$

which is proved in the same way as the inequality above.

Proof of Theorem 5.

Sufficiency. Suppose that for arbitrary $\epsilon > 0$ there exists an elementary set B such that

$$\mu^*(A \Delta B) < \epsilon.$$

Then, according to the Lemma,

(1) $$| \mu^*(A) - m'(B) | = | \mu^*(A) - \mu^*(B) | < \epsilon.$$

In the same way, since

$$(E \setminus A) \Delta (E \setminus B) = A \Delta B,$$

it follows that

(2) $$| \mu^*(E \setminus A) - m'(E \setminus B) | < \epsilon.$$

Inequalities (1) and (2) and

$$m'(B) + m'(E \setminus B) = m'(E) = 1$$

imply that

$$| \mu^*(A) + \mu^*(E \setminus A) - 1 | < 2\epsilon.$$

Since $\epsilon > 0$ is arbitrary,

$$\mu^*(A) + \mu^*(E \setminus A) = 1,$$

and the set A is measurable.

Necessity. Suppose that A is measurable, i.e.,

$$\mu^*(A) + \mu^*(E \setminus A) = 1.$$

For arbitrary $\epsilon > 0$ there exist sets of rectangles $\{B_n\}$ and $\{C_n\}$ such that

$$A \subseteq \bigcup_n B_n, \qquad E \setminus A \subseteq \bigcup_n C_n$$

and such that

$$\sum_n m(B_n) \leq \mu^*(A) + \epsilon/3, \qquad \sum_n m(C_n) \leq \mu^*(E \setminus A) + \epsilon/3.$$

Since $\sum_n m(B_n) < \infty$, there is an N such that

$$\sum_{n>N} m(B_n) < \epsilon/3;$$

set

$$B = \sum_{n=1}^{n=N} B_n.$$

It is clear that the set

$$P = \bigcup_{n>N} B_n$$

contains $A \setminus B$, while the set

$$Q = \bigcup_n (B \cap C_n)$$

contains $B \setminus A$. Consequently, $A \triangle B \subseteq P \cup Q$. Also

$$\mu^*(P) \leq \sum_{n>N} m(B_n) < \epsilon/3.$$

Let us estimate $\mu^*(Q)$. To this end, we note that

$$(\bigcup_n B_n) \cup (\bigcup_n (C_n \setminus B)) = E,$$

and consequently

(3) $$\sum_n m(B_n) + \sum_n m'(C_n \setminus B) \geq 1.$$

But, by hypothesis,

(4) $$\sum_n m(B_n) + \sum_n m(C_n) \leq \mu^*(A) + \mu^*(E \setminus A) + 2\epsilon/3$$
$$= 1 + 2\epsilon/3.$$

From (3) and (4) we obtain

$$\sum_n m(C_n) - \sum_n m'(C_n \setminus B) = \sum_n m'(C_n \cap B) < 2\epsilon/3,$$

.e.,

$$\mu^*(Q) < 2\epsilon/3.$$

Therefore,

$$\mu^*(A \triangle B) \leq \mu^*(P) + \mu^*(Q) < \epsilon.$$

Hence, if A is measurable, for every $\epsilon > 0$ there exists an elementary set B such that $\mu^*(A \triangle B) < \epsilon$. This proves Theorem 5.

THEOREM 6. *The union and intersection of a finite number of measurable sets are measurable sets.*

Proof. It is clearly enough to prove the assertion for two sets. Suppose that A_1 and A_2 are measurable sets. Then for arbitrary $\epsilon > 0$ there are elementary sets B_1 and B_2 such that

$$\mu^*(A_1 \triangle B_1) < \epsilon/2, \qquad \mu^*(A_2 \triangle B_2) < \epsilon/2.$$

Since

$$(A_1 \cup A_2) \triangle (B_1 \cup B_2) \subseteq (A_1 \triangle B_1) \cup (A_2 \triangle B_2),$$

it follows that

(5) $\quad \mu^*[(A_1 \cup A_2) \triangle (B_1 \cup B_2)] \leq \mu^*(A_1 \triangle B_1) + \mu^*(A_2 \triangle B_2) < \epsilon.$

Since $B_1 \cup B_2$ is an elementary set, it follows from Theorem 4 that $A_1 \cup A_2$ is measurable.

But in view of the definition of measurable set, if A is measurable, so is $E \setminus A$; hence, $A_1 \cap A_2$ is measurable because of the relation

$$A_1 \cap A_2 = E \setminus [(E \setminus A_1) \cup (E \setminus A_2)].$$

COROLLARY. *The difference and symmetric difference of two measurable sets are measurable.*

This follows from Theorem 6 and the relations

$$A_1 \setminus A_2 = A_1 \cap (E \setminus A_2),$$
$$A_1 \triangle A_2 = (A_1 \setminus A_2) \cup (A_2 \setminus A_1).$$

THEOREM 7. *If $A_1, \cdots, A_n$ are pairwise disjoint measurable sets, then*

$$\mu(\textstyle\bigcup_{k=1}^{n} A_k) = \sum_{k=1}^{n} \mu(A_k).$$

Proof. As in Theorem 6, it is sufficient to consider the case $n = 2$. Choose an arbitrary $\epsilon > 0$ and elementary sets B_1 and B_2 such that

(6) $\qquad\qquad\qquad \mu^*(A_1 \triangle B_1) < \epsilon,$

(7) $\qquad\qquad\qquad \mu^*(A_2 \triangle B_2) < \epsilon.$

Set $A = A_1 \cup A_2$ and $B = B_1 \cup B_2$. According to Theorem 6, the set A is measurable. Since

$$B_1 \cap B_2 \subseteq (A_1 \triangle B_1) \cup (A_2 \triangle B_2),$$

(8) $\qquad\qquad\qquad m'(B_1 \cap B_2) \leq 2\epsilon.$

In virtue of the Lemma to Theorem 5, (6) and (7) imply that

(9) $\qquad\qquad\qquad | m'(B_1) - \mu^*(A_1) | < \epsilon,$

(10) $| m'(B_2) - \mu^*(A_2) | < \epsilon.$

Since the measure is additive on the class of elementary sets, (8), (9) and (10) yield

$$m'(B) = m'(B_1) + m'(B_2) - m'(B_1 \cap B_2) \geq \mu^*(A_1) + \mu^*(A_2) - 4\epsilon.$$

Noting that $A \triangle B \subseteq (A_1 \triangle B_1) \cup (A_2 \triangle B_2)$, we finally have

$$\mu^*(A) \geq m'(B) - \mu^*(A \triangle B) \geq m'(B) - 2\epsilon \geq \mu^*(A_1) + \mu^*(A_2) - 6\epsilon.$$

Since 6ϵ may be chosen arbitrarily small,

$$\mu^*(A) \geq \mu^*(A_1) + \mu^*(A_2).$$

Inasmuch as the converse inequality

$$\mu^*(A) \leq \mu^*(A_1) + \mu^*(A_2)$$

is always true for $A = A_1 \cup A_2$, we have

$$\mu^*(A) = \mu^*(A_1) + \mu^*(A_2).$$

Since A_1, A_2 and A are measurable, μ^* can be replaced by μ, and this proves the theorem.

THEOREM 8. *The union and intersection of a countable number of measurable sets are measurable sets.*

Proof. Let

$$A_1, \cdots, A_n, \cdots$$

be a countable collection of measurable sets, and let $A = \bigcup_{n=1}^{\infty} A_n$. Set $A_n' = A_n \setminus \bigcup_{k=1}^{n-1} A_k$. It is clear that $A = \bigcup_{n=1}^{\infty} A_n'$ and that the sets A_n' are pairwise disjoint. By Theorem 6 and its Corollary, all the sets A_n' are measurable. According to Theorems 7 and 3,

$$\sum_{k=1}^{n} \mu(A_k') = \mu(\bigcup_{k=1}^{n} A_k') \leq \mu(A)$$

for arbitrary finite n. Therefore, the series

$$\sum_{n=1}^{\infty} \mu(A_n')$$

converges, and consequently for arbitrary $\epsilon > 0$ there exists an N such that

(11) $\sum_{n>N} \mu(A_n') < \epsilon/2.$

Since the set $C = \bigcup_{n=1}^{N} A_n'$ is measurable (as a union of a finite number of measurable sets), there exists an elementary set B such that

(12) $\mu^*(C \triangle B) < \epsilon/2.$

Inasmuch as

$$A \triangle B \subseteq (C \triangle B) \cup (\bigcup_{n>N} A_n'),$$

(11) and (12) imply that

$$\mu^*(A \bigtriangleup B) < \epsilon.$$

Hence, by Theorem 5, the set A is measurable.

Since the complement of a measurable set is measurable, the second half of the theorem follows from the relation

$$\bigcap_n A_n = E \setminus \bigcup_n (E \setminus A_n).$$

Theorem 8 is a generalization of Theorem 6. The following theorem is the corresponding generalization of Theorem 7.

THEOREM 9. *If* $\{A_n\}$ *is a sequence of pairwise disjoint measurable sets, and* $A = \bigcup_n A_n$, *then*

$$\mu(A) = \sum_n \mu(A_n).$$

Proof. By Theorem 7, for arbitrary N

$$\mu(\bigcup_{n=1}^{N} A_n) = \sum_{n=1}^{N} \mu(A_n) \leq \mu(A).$$

Letting $N \to \infty$, we obtain

(13) $$\mu(A) \geq \sum_{n=1}^{\infty} \mu(A_n).$$

On the other hand, according to Theorem 3,

(14) $$\mu(A) \leq \sum_{n=1}^{\infty} \mu(A_n).$$

The theorem follows from (13) and (14).

The property of the measure established in Theorem 9 is called *complete additivity* or *σ-additivity*. The following property of the measure, called *continuity*, is an immediate consequence of σ-additivity.

THEOREM 10. *If* $A_1 \supseteq A_2 \supseteq \cdots$ *is a monotone decreasing sequence of measurable sets, and* $A = \bigcap_n A_n$, *then*

$$\mu(A) = \lim_{n \to \infty} \mu(A_n).$$

It is obviously sufficient to consider the case $A = \emptyset$, since the general case reduces to this on replacing A_n by $A_n \setminus A$. Then

$$A_1 = (A_1 \setminus A_2) \cup (A_2 \setminus A_3) \cup \cdots$$

and

$$A_n = (A_n \setminus A_{n+1}) \cup (A_{n+1} \setminus A_{n+2}) \cup \cdots.$$

Consequently,

(15) $$\mu(A_1) = \sum_{k=1}^{\infty} \mu(A_k \setminus A_{k+1})$$

and

(16) $$\mu(A_n) = \sum_{k=n}^{\infty} \mu(A_k \setminus A_{k+1});$$

since the series (15) converges, its remainder (16) approaches zero as $n \to \infty$. Hence,

$$\mu(A_n) \to 0 \qquad\qquad (n \to \infty).$$

This is what we were to prove.

COROLLARY. *If $A_1 \subseteq A_2 \subseteq \cdots$ is a monotone increasing sequence of measurable sets and $A = \bigcup_n A_n$, then*

$$\mu(A) = \lim_{n \to \infty} \mu(A_n).$$

To prove this it is sufficient to replace the sets A_n by their complements and then to use Theorem 10.

We have now extended the measure defined on the elementary sets to the wider class of measurable sets. The latter class is closed with respect to the operations of countable unions and intersections, and the measure on this class is σ-additive.

We conclude this section with a few remarks.

1. The theorems we have proved characterize the class of Lebesgue measurable sets.

Since every open set contained in E can be written as a union of a countable number of open rectangles, that is, measurable sets, Theorem 8 implies that every open set is measurable. The closed sets are also measurable, since they are the complements of the open sets. In view of Theorem 8, all sets which can be obtained from the open and closed sets by taking countable unions and intersections are also measurable. It can be shown, however, that these sets do not exhaust the class of all Lebesgue measurable sets.

2. We have considered only plane sets contained in the unit square $E = \{0 \le x, y \le 1\}$. It is not hard to remove this restriction. This can be done, for instance, in the following way. Representing the whole plane as the union of the squares $E_{nm} = \{n \le x \le n + 1, m \le y \le m + 1 \ (m, n \text{ integers})\}$, we define a plane set A to be measurable if its intersection $A_{nm} = A \cap E_{nm}$ with each of these squares is measurable, and the series

$$\sum_{n,m} \mu(A_{nm})$$

converges. We then define

$$\mu(A) = \sum_{n,m} \mu(A_{nm}).$$

All the measure properties derived above carry over in an obvious fashion to this case.

3. In this section we have constructed Lebesgue measure for plane sets. Lebesgue measure on the line, in three dimensions, or, in general, in Euclidean n-space, can be constructed analogously. The measure in all these

cases is constructed in the same way: starting with a measure defined for a certain class of simple sets (rectangles in the plane; open, closed and half-open intervals on the line; etc.) we first define a measure for finite unions of such sets, and then extend it to the much wider class of Lebesgue measurable sets. The definition of measurable set is carried over verbatim to sets in a space of arbitrary (finite) dimension.

4. To introduce Lebesgue measure we started with the usual definition of area. The analogous construction in one dimension is based on the length of an interval. However, the concept of measure can be introduced in another, somewhat more general, way.

Let $F(t)$ be a nondecreasing and left continuous function defined on the real line. We set

$$m(a, b) = F(b) - F(a + 0),$$
$$m[a, b] = F(b + 0) - F(a),$$
$$m(a, b] = F(b + 0) - F(a + 0),$$
$$m[a, b) = F(b) - F(a).$$

It is easily verified that the interval function m defined in this way is nonnegative and additive. Proceeding in the same way as described above, we can construct a certain "measure" $\mu_F(A)$. The class of sets measurable relative to this measure is closed under the operations of countable unions and intersections, and μ_F is σ-additive. The class of μ_F-measurable sets will, in general, depend on the choice of the function F. However, the open and closed sets, and consequently their countable unions and intersections, will be measurable for arbitrary choice of F. The measures μ_F, where F is arbitrary (except for the conditions imposed above), are called Lebesgue-Stieltjes measures. In particular, the function $F(t) = t$ corresponds to the usual Lebesgue measure on the real line.

A measure μ_F which is equal to zero on every set whose Lebesgue measure is zero is said to be *absolutely continuous*. A measure μ_F whose set of values is countable [this will occur whenever the set of values of $F(t)$ is countable] is said to be *discrete*. A measure μ_F is called *singular* if it is zero on every set consisting of one point, and if there is a set M whose Lebesgue measure is zero and such that the μ_F measure of its complement is zero.

It can be proved that every measure μ_F is a sum of an absolutely continuous, a discrete and a singular measure.

★ *Existence of nonmeasurable sets.* We proved above that the class of Lebesgue measurable sets is very wide. The question naturally arises whether there exist nonmeasurable sets. We shall prove that there are such sets. The simplest example of a nonmeasurable set can be constructed on a circumference.

Let C be a circumference of length 1, and let α be an irrational number. Partition the points of C into classes by the following rule: Two points of C belong to the same class if, and only if, one can be carried into the other by a rotation of C through an angle $n\alpha$ (n an integer). Each class is clearly countable. We now select a point from each class. We show that the resulting set Φ is nonmeasurable. Denote by Φ_n the set obtained by rotating Φ through the angle $n\alpha$. It is easily seen that all the sets Φ_n are pairwise disjoint and that their union is C. If the set Φ were measurable, the sets Φ_n congruent to it would also be measurable. Since

$$C = \bigcup_{n=-\infty}^{\infty} \Phi_n , \qquad \Phi_n \cap \Phi_m = \emptyset \qquad\qquad (n \neq m),$$

the σ-additivity of the measure would imply that

(17) $$\sum_{n=-\infty}^{\infty} \mu(\Phi_n) = 1.$$

But congruent sets must have the same measure:

$$\mu(\Phi_n) = \mu(\Phi).$$

The last equality shows that (17) is impossible, since the sum of the series on the left side of (17) is zero if $\mu(\Phi) = 0$, and is infinity if $\mu(\Phi) > 0$. Hence, the set Φ (and consequently every set Φ_n) is nonmeasurable. $\star$

EXERCISES

1. If A is a countable set of points contained in

$$E = \{(x, y) : 0 \leq x \leq 1, 0 \leq y \leq 1\},$$

then A is measurable and $\mu(A) = 0$.

2. Let $F_0 = [0, 1]$ and let F be the Cantor set constructed on F_0 (see vol. 1, pp. 32–33). Prove that $\mu_1(F) = 0$, where $\mu_1(F)$ is the (linear) Lebesgue measure of F.

3. Let F be as in Ex. 2. If $x \in F$, then

$$x = a_1/3 + \cdots + a_n/3^n + \cdots,$$

where $a_1 = 0$ or 2. Define

$$\varphi(x) = a_1/2^2 + \cdots + a_n/2^{n+1} + \cdots \qquad (x \in F)$$

(see the reference given in Ex. 2). The function φ is single-valued. If a, $b \in F$ are such that $(a, b) \notin F$ [i.e., (a, b) is a deleted open interval in the construction of F], show that $\varphi(a) = \varphi(b)$. We can therefore define φ on $[a, b]$ as equal to this common value. The function φ so defined on $F_0 = [0, 1]$ is nondecreasing and continuous. Show that μ_φ, the Lebesgue-Stieltjes measure generated by φ on the set F_0, is a singular measure. The function φ is called the Cantor function.

4. For $E = \{(x, y) : 0 \leq x \leq 1, 0 \leq y \leq 1\}$, $A \subset E$ we can restate our definition for the measurability of A as follows: A is measurable provided

$$\mu^*(E) = \mu^*(E \cap A) + \mu^*(E \setminus A).$$

Show that A satisfies the measurability criterion of Carathéodory: For every $F \subseteq E$,

$$\mu^*(F) = \mu^*(F \cap A) + \mu^*(F \setminus A).$$

The converse implication is, of course, trivial.

5. Lebesgue measure in the plane is *regular*, i.e.,

$$\mu^*(A) = \inf \{\mu(G) : A \subseteq G, G \text{ open relative to } E\}.$$

6. Derive Lebesgue's criterion for measurability: A set $A \subseteq E$ is measurable if, and only if, for every $\epsilon > 0$ there exist G open (relative to E) and F closed such that $F \subset A \subset G$ and $\mu(G \setminus F) < \epsilon$. (See the definition of Jordan measurability in §36.) Hint: Apply Ex. 5 to A and $E \setminus A$.

§34. Collections of sets

Our discussion of the abstract theory of measure will presuppose certain facts about collections of sets, in addition to the elementary theory of sets presented in Chapter I.

A *collection of sets* is a set whose elements are themselves sets. As a rule, we shall consider collections of sets whose elements are subsets of a fixed set X. In general, collections of sets will be denoted by capital German letters. Fundamentally, we shall be interested in collections of sets which are closed under some (or all) of the operations introduced in Chapter I, §1.

DEFINITION 1. A *ring* is a nonempty collection of sets $\mathfrak{R}$ with the property that $A \in \mathfrak{R}$, $B \in \mathfrak{R}$ imply that $A \triangle B \in \mathfrak{R}$ and $A \cap B \in \mathfrak{R}$.

Since

$$A \cup B = (A \triangle B) \triangle (A \cap B),$$

$$A \setminus B = A \triangle (A \cap B)$$

for arbitrary A and B, it follows that $A \in \mathfrak{R}$, $B \in \mathfrak{R}$ imply that $A \cup B \in \mathfrak{R}$ and $A \setminus B \in \mathfrak{R}$. Hence a ring of sets is a collection of sets closed under unions, intersections, differences and symmetric differences (of pairs of sets). Clearly, a ring is also closed under finite unions and intersections:

$$C = \bigcup_{k=1}^{n} A_k, \qquad D = \bigcap_{k=1}^{n} A_k.$$

Every ring contains the empty set $\emptyset$, since $A \setminus A = \emptyset$. A ring consisting of the empty set alone is the smallest possible ring.

A set E is called a *unit* of a collection of sets $\mathfrak{S}$ if it is an element of $\mathfrak{S}$ and if

$$A \cap E = A$$

for arbitrary $A \in \mathfrak{S}$. It is easily seen that if $\mathfrak{S}$ has a unit, it is unique.

Hence, the unit of a collection of sets $\mathfrak{S}$ is the maximal set of the collection, that is, the set which contains every other element of $\mathfrak{S}$.

A ring of sets with a unit is called an *algebra* of sets. [TRANS. NOTE. This definition leads to difficulties in the statements and proofs of certain theorems in the sequel. These difficulties disappear if the usual definition of an algebra is used: Let X be a set, $\mathfrak{S}$ a collection of subsets of X. The collection $\mathfrak{S}$ is called an *algebra* if $\mathfrak{S}$ is a ring with unit $E = X$.]

EXAMPLES. 1. If A is an arbitrary set, the collection $\mathfrak{M}(A)$ of all its subsets is an algebra of sets with unit $E = A$.

2. If A is an arbitrary nonempty set, the collection $\{\emptyset, A\}$ consisting of the set A and the empty set $\emptyset$ is an algebra with unit $E = A$.

3. The set of all finite subsets of an arbitrary set A is a ring. This ring is an algebra if, and only if, A is finite.

4. The set of all bounded subsets of the real line is a ring without a unit.

An immediate consequence of the definition of a ring is

THEOREM 1. *The intersection* $\mathfrak{R} = \bigcap_\alpha \mathfrak{R}_\alpha$ *of an arbitrary number of rings is a ring.*

We shall prove the following simple, but important, proposition:

THEOREM 2. *If* $\mathfrak{S}$ *is an arbitrary nonempty collection of sets, there exists precisely one ring* $\mathfrak{R}(\mathfrak{S})$ *containing* $\mathfrak{S}$ *and contained in every ring* $\mathfrak{R}$ *containing* $\mathfrak{S}$.

Proof. It is easy to see that the ring $\mathfrak{R}(\mathfrak{S})$ is uniquely determined by $\mathfrak{S}$. To show that it exists, we consider the union $X = \bigcup_{A \in \mathfrak{S}} A$ and the ring $\mathfrak{M}(X)$ of all the subsets of X. Let Σ be the collection of all rings contained in $\mathfrak{M}(X)$ and containing $\mathfrak{S}$. The intersection

$$\mathfrak{P} = \bigcap_{\mathfrak{R} \in \Sigma} \mathfrak{R}$$

is obviously the required ring $\mathfrak{R}(\mathfrak{S})$.

For, if $\mathfrak{R}^*$ is a ring containing $\mathfrak{S}$, then $\mathfrak{R} = \mathfrak{R}^* \cap \mathfrak{M}(X)$ is a ring in Σ; hence,

$$\mathfrak{S} \subseteq \mathfrak{P} \subseteq \mathfrak{R} \subseteq \mathfrak{R}^*,$$

that is, $\mathfrak{P}$ is minimal. $\mathfrak{R}(\mathfrak{S})$ is called the *minimal ring over the collection* $\mathfrak{S}$. [$\mathfrak{R}(\mathfrak{S})$ is also called the *ring generated by* $\mathfrak{S}$.]

The actual construction of the ring $\mathfrak{R}(\mathfrak{S})$ over a prescribed collection $\mathfrak{S}$ is, in general, quite complicated. However, it becomes completely explicit in the important special case when $\mathfrak{S}$ is a semi-ring.

DEFINITION 2. A collection of sets $\mathfrak{S}$ is called a *semi-ring* if it satisfies the following conditions:

(1) $\mathfrak{S}$ contains the empty set $\emptyset$.

(2) If $A, B \in \mathfrak{S}$, then $A \cap B \in \mathfrak{S}$.

(3) If A and $A_1 \subseteq A$ are both elements of $\mathfrak{S}$, then

$$A = \bigcup_{k=1}^{n} A_k,$$

where the sets A_k are pairwise disjoint elements of $\mathfrak{S}$, and the first of the sets A_k is the given set A_1.

In the sequel we shall call a collection of pairwise disjoint sets

$$A_1, \cdots, A_n,$$

whose union is a set A, a *finite partition* of the set A.

Every ring $\mathfrak{R}$ is a semi-ring, since if both A and $A_1 \subseteq A$ belong to $\mathfrak{R}$, then $A = A_1 \cup A_2$, where $A_2 = A \setminus A_1 \in \mathfrak{R}$.

An example of a semi-ring which is not a ring is the collection of all open, closed and half-open intervals on the real line. [Among the intervals we include, of course, the empty interval (a, a) and the interval consisting of one point $[a, a]$.]

In order to show how the minimal ring over a semi-ring is constructed, we derive several properties of semi-rings.

LEMMA 1. *Suppose that $A_1, \cdots, A_n, A$ are all elements of a semi-ring $\mathfrak{S}$, where the sets A_i are pairwise disjoint subsets of A. Then there is a finite partition of A:*

$$A = \bigcup_{k=1}^{s} A_k \qquad (s \geq n, A_k \in \mathfrak{S}),$$

whose first n terms are the sets A_i $(1 \leq i \leq n)$.

The proof is by induction. The assertion is true for $n = 1$ by the definition of a semi-ring. We assume that the proposition is true for $n = m$ and consider $m + 1$ sets $A_1, \cdots, A_m, A_{m+1}$ satisfying the hypothesis of the lemma. In view of the inductive hypothesis,

$$A = A_1 \cup A_2 \cup \cdots \cup A_m \cup B_1 \cup B_2 \cup \cdots \cup B_p,$$

where all the sets B_q $(1 \leq q \leq p)$ are elements of $\mathfrak{S}$. Set

$$B_{q1} = A_{m+1} \cap B_q.$$

By the definition of a semi-ring there is a partition

$$B_q = B_{q1} \cup B_{q2} \cup \cdots \cup B_{qr_q},$$

where all the sets B_{qj} are elements of $\mathfrak{S}$. It is easy to see that

$$A = A_1 \cup \cdots \cup A_m \cup A_{m+1} \cup \bigcup_{q=1}^{p} \bigcup_{j=2}^{r_q} B_{qj}.$$

Hence, the lemma is true for $n = m + 1$, and so for all n.

LEMMA 2. *If $A_1, \cdots, A_n$ are elements of a semi-ring $\mathfrak{S}$, there exists in $\mathfrak{S}$ a finite set of pairwise disjoint sets $B_1, \cdots, B_t$ such that each A_k can be written as a union*

$$A_k = \bigcup_{s \in M_k} B_s$$

of some of the sets B_s.

Proof. The lemma is trivial for $n = 1$, since it is then enough to put $t = 1$, $B_1 = A_1$. Suppose that the lemma is true for $n = m$ and consider a collection of sets $A_1, \cdots, A_{m+1}$. Let $B_1, \cdots, B_t$ be elements of $\mathfrak{S}$ satisfying the conditions of the lemma relative to the sets $A_1, \cdots, A_m$. Set

$$B_{s1} = A_{m+1} \cap B_s.$$

By Lemma 1, there exists a partition

$$(1) \qquad A_{m+1} = \bigcup_{s=1}^{t} B_{s1} \cup \bigcup_{p=1}^{q} B_p' \qquad (B_p' \in \mathfrak{S}),$$

and in view of the definition of a semi-ring there exists a partition

$$B_s = B_{s1} \cup B_{s2} \cup \cdots \cup B_{sf_s} \qquad (B_{sq} \in \mathfrak{S}).$$

It is easily seen that

$$A_k = \bigcup_{s \in M_k} \bigcup_{q=1}^{f_s} B_{sq} \qquad (1 \le k \le m),$$

and that the sets B_{sq}, B_p' are pairwise disjoint. Hence, the sets B_{sq}, B_p' satisfy the lemma relative to the sets $A_1, \cdots, A_m, A_{m+1}$. This proves the lemma.

LEMMA 3. *If $\mathfrak{S}$ is a semi-ring, then $\Re(\mathfrak{S})$ coincides with the collection $\mathfrak{Z}$ of the sets A which admit of a finite partition*

$$A = \bigcup_{k=1}^{n} A_k \qquad (A_k \in \mathfrak{S}).$$

Proof. We show that $\mathfrak{Z}$ is a ring. If $A, B \in \mathfrak{Z}$, then

$$A = \bigcup_{k=1}^{n} A_k, \qquad B = \bigcup_{k=1}^{m} B_k \qquad (A_k, B_k \in \mathfrak{S}).$$

Since $\mathfrak{S}$ is a semi-ring, the sets

$$C_{ij} = A_i \cap B_j$$

are also elements of $\mathfrak{S}$. By Lemma 1,

$$(2) \qquad A_i = \bigcup_j C_{ij} \cup \bigcup_{k=1}^{r_i} D_{ik}; \qquad B_j = \bigcup_i C_{ij} \cup \bigcup_{k=1}^{s_j} E_{jk},$$

where $D_{ik}, E_{jk} \in \mathfrak{S}$. The equality (2) implies that

$$A \cap B = \bigcup_{i,j} C_{ij},$$

$$A \bigtriangleup B = \bigcup_{i,k} D_{ik} \cup \bigcup_{j,k} E_{jk}.$$

Therefore, $A \cap B$ and $A \triangle B$ are elements of $\mathcal{B}$. Hence $\mathcal{B}$ is a ring, and it is obvious that it is the minimal ring containing $\mathfrak{S}$.

In various problems, especially in measure theory, it is necessary to consider denumerable, as well as finite, unions and intersections. It is therefore necessary to introduce, in addition to the definition of a ring, the following definitions.

DEFINITION 3. A ring $\mathfrak{R}$ of sets is called a *σ-ring* if $A_i \in \mathfrak{R}$ $(i = 1, 2, \cdots)$ implies that

$$S = \bigcup_n A_n \in \mathfrak{R}.$$

DEFINITION 4. A ring of sets $\mathfrak{R}$ is called a *δ-ring* if $A_i \in \mathfrak{R}$ $(i = 1, 2, \cdots)$ implies that

$$D = \bigcap_n A_n \in \mathfrak{R}.$$

It is natural to call a σ-ring (δ-ring) with a unit a σ-algebra (δ-algebra). However, it is easy to see that these two notions coincide: every σ-algebra is a δ-algebra and every δ-algebra is a σ-algebra. This follows from de Morgan's laws:

$$\bigcup_n A_n = E \setminus \bigcap_n (E \setminus A_n),$$
$$\bigcap_n A_n = E \setminus \bigcup_n (E \setminus A_n)$$

(see Chapter 1, §1). σ-algebras, or δ-algebras, are called *Borel algebras*; or, briefly, *B-algebras*.

The simplest example of a *B*-algebra is the collection of all subsets of a set A.

For *B*-algebras there is a theorem analogous to Theorem 2, which was proved above for rings.

THEOREM 4. *If $\mathfrak{S}$ is a nonempty collection of sets, there exists a B-algebra $\mathfrak{B}(\mathfrak{S})$ containing $\mathfrak{S}$ and contained in every B-algebra containing $\mathfrak{S}$.*

The proof (see Trans. Note, p. 16) is carried out in exactly the same way as the proof of Theorem 2. The *B*-algebra $\mathfrak{B}(\mathfrak{S})$ is called the *minimal B-algebra over the system* $\mathfrak{S}$ or the *Borel closure* of $\mathfrak{S}$.

In analysis an important part is played by the *Borel sets* or *B-sets*, which may be defined as the elements of the minimal *B*-algebra over the set of all closed intervals $[a, b]$ on the real line (or the set of all open intervals, or the set of half-closed intervals).

To supplement §7 of Chapter 1 we note the following facts, which will be required in Chapter VI.

Let $y = f(x)$ be a function defined on a set M with values in a set N. Denote by $f(\mathfrak{M})$ the collection of all images $f(A)$ of sets in $\mathfrak{M}$, where $\mathfrak{M}$ is a set of subsets of M. Similarly, let $f^{-1}(\mathfrak{N})$ be the collection of all inverse images $f^{-1}(A)$, where $\mathfrak{N}$ is a set of subsets of N. Then:

1. If $\Re$ is a ring, $f^{-1}(\Re)$ is a ring.
2. If $\Re$ is an algebra, $f^{-1}(\Re)$ is an algebra.
3. If $\Re$ is a B-algebra, $f^{-1}(\Re)$ is a B-algebra.
4. $\Re(f^{-1}(\Re)) = f^{-1}(\Re(\Re))$.
5. $\mathfrak{B}(f^{-1}(\Re)) = f^{-1}(\mathfrak{B}(\Re))$.

$\star$ Let $\Re$ be a ring of sets. If the operations $A \triangle B$ and $A \cap B$ are regarded as addition and multiplication, respectively, then $\Re$ is a ring in the usual algebraic sense. All its elements satisfy the conditions

(*) $$a + a = 0, \qquad a^2 = a.$$

A ring all of whose elements satisfy the conditions (*) is called a Boolean ring. Every Boolean ring can be realized as a ring of sets with the operations $A \triangle B$ and $A \cap B$ (Stone). $\star$

EXERCISES

1. Suppose that $\Re$ is a ring of subsets of a set X and that $\mathfrak{A}$ is the collection of those sets $E \subseteq X$ for which either $E \in \Re$ or else $X \setminus E \in \Re$. Show that $\mathfrak{A}$ is an algebra with unit X.

2. Determine the minimal ring $\Re(\mathfrak{S})$ in each of the following cases:
   (a) for a fixed subset $A \subseteq X$, $\mathfrak{S} = \{A\}$;
   (b) for a fixed subset $A \subseteq X$, $\mathfrak{S} = \{B : A \subseteq B \subseteq X\}$.

3. Let $\mathfrak{S}$ be a semi-ring in X, and let $\Re(\mathfrak{S})$ be the minimal ring over $\mathfrak{S}$. Then the minimal σ-rings over $\mathfrak{S}$ and $\Re(\mathfrak{S})$ coincide.

4. For each of the following sets what are the σ-ring and the Borel algebra generated by the given class of sets $\mathfrak{S}$?
   (a) Let T be a one-to-one onto transformation of X with itself. A subset $A \subseteq X$ is called invariant if $x \in A$ implies that $T(x) \in A$ and $T^{-1}(x) \in A$. Let $\mathfrak{S}$ be the collection of invariant subsets of X.
   (b) Let X be the plane and let $\mathfrak{S}$ be the collection of all subsets of the plane which can be covered by countably many horizontal lines.

§35. Measures on semi-rings. Extension of a measure on a semi-ring to the minimal ring over the semi-ring

In §33, to define a measure in the plane we started with the measure (area) of rectangles and then extended this measure to a more general class of sets. The results and methods of §33 are completely general and can be extended, with no essential changes, to measures defined on arbitrary sets. The first step in the construction of a measure in the plane is the extension of the measure of rectangles to elementary sets, that is, to finite unions of pairwise disjoint rectangles.

We consider the abstract analogue of this problem in this section.

DEFINITION 1. A set function $\mu(A)$ is called a *measure* if

1) its domain of definition S_μ is a semi-ring;
2) its values are real and nonnegative;
3) it is additive, that is, if

$$A = \bigcup_k A_k$$

is a finite partition of a set $A \in S_\mu$ in sets $A_k \in S_\mu$, then

$$\mu(A) = \sum_k \mu(A_k).$$

REMARK. Since $\emptyset = \emptyset \cup \emptyset$, it follows that $\mu(\emptyset) = 2\mu(\emptyset)$, i.e., $\mu(\emptyset) = 0$.
The following two theorems on measures in semi-rings will be applied repeatedly in the sequel.

THEOREM 1. *Let μ be a measure defined on a semi-ring S_μ. If*

$$A_1, \cdots, A_n, A \in S_\mu,$$

where the sets A_k are pairwise disjoint subsets of A, then

$$\sum_{k=1}^n \mu(A_k) \leq \mu(A).$$

Proof. Since S_μ is a semi-ring, in view of Lemma 1 of §34 there exists a partition

$$A = \bigcup_{k=1}^s A_k \qquad (s \geq n, A_k \in S_\mu)$$

in which the first n sets coincide with the given sets $A_1, \cdots, A_n$. Since the measure of an arbitrary set is nonnegative,

$$\sum_{k=1}^n \mu(A_k) \leq \sum_{k=1}^s \mu(A_k) = \mu(A).$$

THEOREM 2. *If $A_1, \cdots, A_n, A \in S_\mu$ and $A \subseteq \bigcup_{k=1}^n A_k$, then*

$$\mu(A) \leq \sum_{k=1}^n \mu(A_k).$$

Proof. According to Lemma 2 of §34 there exist pairwise disjoint sets $B_1, \cdots, B_t \in S_\mu$ such that each of the sets $A, A_1, \cdots, A_n$ can be written as a union of some of the sets B_s :

$$A = \bigcup_{s \in M_0} B_s ; \qquad A_k = \bigcup_{s \in M_k} B_s \qquad (1 \leq k \leq n)$$

where each index $s \in M_0$ is an element of some M_k. Consequently, every term of the sum

$$\sum_{s \in M_0} \mu(B_s) = \mu(A)$$

appears at least once in the double sum

$$\sum_{k=1}^n \sum_{s \in M_k} \mu(B_s) = \sum_{k=1}^n \mu(A_k).$$

Hence,

$$\mu(A) \leq \sum_{k=1}^n \mu(A_k).$$

In particular, if $n = 1$, we obtain the

COROLLARY. *If $A \subseteq A'$, then $\mu(A) \leq \mu(A')$.*

DEFINITION 2. A measure $\mu(A)$ is said to be an *extension of a measure* $m(A)$ if $S_m \subseteq S_\mu$ and if $\mu(A) = m(A)$ for every $A \in S_m$.

The primary purpose of this section is to prove the following theorem.

THEOREM 3. *Every measure $m(A)$ has a unique extension $\mu(A)$ whose domain of definition is the ring $\Re(S_m)$.*

Proof. For each set $A \in \Re(S_m)$ there exists a partition

$$(1) \qquad\qquad A = \bigcup_{k=1}^{n} B_k \qquad\qquad (B_k \in S_m)$$

(§34, Theorem 3). We set, by definition,

$$(2) \qquad\qquad \mu(A) = \sum_{k=1}^{n} m(B_k).$$

It is easily seen that the value of $\mu(A)$ defined by (2) is independent of the choice of the partition (1). In fact, let

$$A = \bigcup_{i=1}^{m} B_i = \bigcup_{j=1}^{n} C_j \qquad (B_i \in S_m, \, C_j \in S_m)$$

be two partitions of A. Since all the intersections $B_i \cap C_j$ belong to S_m, in view of the additivity of the measure m,

$$\sum_i m(B_i) = \sum_{i=1}^{m} \sum_{j=1}^{n} m(B_i \cap C_j) = \sum_j m(C_j).$$

The measure $\mu(A)$ defined by (2) is obviously nonnegative and additive. This proves the existence of an extension $\mu(A)$ of the measure m. To prove ts uniqueness, we note that, according to the definition of an extension, if $A = \bigcup_{k=1}^{n} B_k$, where the B_k are disjoint elements of S_m, then

$$\mu^*(A) = \sum \mu^*(B_k) = \sum m(B_k) = \mu(A)$$

for an arbitrary extension μ^* of m over $\Re(S_m)$. This proves the theorem.

The relation of this theorem to the constructions of §33 will be fully clear if we note that the set of all rectangles in the plane is a semi-ring, that the area of the rectangles is a measure in the sense of Def. 1 and that the class of elementary plane sets is the minimal ring over the semi-ring of rectangles.

EXERCISES

1. Let X be the set of positive integers, $\mathfrak{S}$ the set of all finite subsets of X. Suppose that $\sum_{n=1}^{\infty} u_n$ is a convergent series of positive numbers. For $A \in \mathfrak{S}$ define $\mu(A) = \sum u_n \, (n \in A)$. Prove that μ is a measure. Now suppose that $\mathfrak{P}$ is the set of all subsets of X and that μ is defined as above for finite subsets of X, but that $\mu(A) = +\infty$ if A is infinite. μ is still finitely additive [although $\mu(A)$ may equal $+\infty$ for some sets]. However, μ is not completely additive (see §37, Def. 1).

2 Let X be the plane, and let $\mathfrak{S} = \{A : A = $ the set of all (x, y) such that $a < x \leq b,\ y = c\}$, i.e., $\mathfrak{S}$ consists of all the horizontal right half-closed line segments. Define $\mu(A) = b - a$.

  (a) Show that $\mathfrak{S}$ is a semi-ring.

  (b) Show that μ is a measure on $\mathfrak{S}$.

3. Let μ be a measure on a ring $\mathfrak{R}$.

  (a) For $A, B \in \mathfrak{R}$ show that $\mu(A \cup B) = \mu(A) + \mu(B) - \mu(A \cap B)$.

  (b) For $A, B, C \in \mathfrak{R}$ show that

$$\mu(A \cup B \cup C) = \mu(A) + \mu(B) + \mu(C)$$
$$- [\mu(A \cap B) + \mu(B \cap C) + \mu(C \cap A)] + \mu(A \cap B \cap C).$$

  (c) Generalize to the case $A_1, \cdots, A_n \in \mathfrak{R}$.

§36. Extension of the Jordan measure

The concept of Jordan measure is of historical and practical interest, but will not be used in the sequel.

In this section we shall consider the general form of the process which in the case of plane figures is used to pass from the definition of the areas of finite unions of rectangles with sides parallel to the coordinate axes to the areas of those figures which are assigned definite areas in elementary geometry or classical analysis. This transition was described with complete clarity by the French mathematician Jordan about 1880. Jordan's basic idea, however, goes back to the mathematicians of ancient Greece and consists in approximating the "measurable" sets A by sets A' and A'' such that

$$A' \subseteq A \subseteq A''.$$

Since an arbitrary measure can be extended to a ring (§35, Theorem 3) it is natural to assume that the initial measure m is defined on the ring $\mathfrak{R} = \mathfrak{R}(S_m)$. We shall make this assumption in the rest of this section.

DEFINITION 1. We shall say that a set A is *Jordan measurable* if for every $\epsilon > 0$ there are sets $A', A'' \in \mathfrak{R}$ such that

$$A' \subseteq A \subseteq A'',\qquad m(A'' \setminus A') < \epsilon.$$

THEOREM 1. *The collection $\mathfrak{R}^*$ of sets which are Jordan measurable is a ring.*

For, suppose that $A, B \in \mathfrak{R}^*$; then for arbitrary $\epsilon > 0$ there exist sets $A', A'', B', B'' \in \mathfrak{R}$ such that

$$A' \subseteq A \subseteq A'',\qquad B' \subseteq B \subseteq B''$$

and

$$m(A'' \setminus A') < \epsilon/2, \qquad m(B'' \setminus B') < \epsilon/2.$$

Hence

(1) $$A' \cup B' \subseteq A \cup B \subseteq A'' \cup B'',$$

(2) $$A' \setminus B'' \subseteq A \setminus B \subseteq A'' \setminus B'.$$

Since

$$(A'' \cup B'') \setminus (A' \cup B') \subseteq (A'' \setminus A') \cup (B'' \setminus B'),$$

(3)
$$\begin{aligned}
m[(A'' \cup B'') \setminus (A' \cup B')] &\leq m[(A'' \setminus A') \cup (B'' \setminus B')] \\
&\leq m(A'' \setminus A') + m(B'' \setminus B') \\
&< \epsilon/2 + \epsilon/2 = \epsilon.
\end{aligned}$$

Since

$$(A'' \setminus B') \setminus (A' \setminus B'') \subseteq (A'' \setminus A') \cup (B'' \setminus B'),$$

(4)
$$\begin{aligned}
m[(A'' \setminus B') \setminus (A' \setminus B'')] &\leq m(A'' \setminus A') \cup (B'' \setminus B')] \\
&\leq m(A'' \setminus A') + m(B'' \setminus B') \\
&< \epsilon/2 + \epsilon/2 = \epsilon.
\end{aligned}$$

Inasmuch as $\epsilon > 0$ is arbitrary and the sets $A' \cup B'$, $A'' \cup B''$, $A' \setminus B''$, $A'' \setminus B'$ are elements of $\mathfrak{R}$, (1), (2), (3) and (4) imply that $A \cup B$ and $A \setminus B$ are elements of $\mathfrak{R}^*$.

Let $\mathfrak{M}$ be the collection consisting of the sets A for which there is a set $B \in \mathfrak{R}$ such that $B \supseteq A$. For arbitrary $A \in \mathfrak{M}$ we define

$$\bar{\mu}(A) = \inf \{m(B); B \supseteq A\},$$

$$\mu(A) = \sup \{m(B); B \subseteq A\}.$$

The functions $\bar{\mu}(A)$ and $\mu(A)$ are called the *outer* and *inner* measures, respectively, of the set A.

Obviously,

$$\mu(A) \leq \bar{\mu}(A).$$

THEOREM 2. *The ring $\mathfrak{R}^*$ coincides with the system of all sets $A \in \mathfrak{M}$ such that $\mu(A) = \bar{\mu}(A)$.*

Proof. If

$$\bar{\mu}(A) \neq \mu(A),$$

then

$$\bar{\mu}(A) - \mu(A) = h > 0,$$

and

$$m(A') \leq \mu(A), \qquad m(A'') \geq \bar{\mu}(A),$$
$$m(A'' \smallsetminus A') = m(A'') - m(A') \geq h > 0$$

for arbitrary A', $A'' \in \Re$ such that $A' \subseteq A \subseteq A''$. Hence $A \notin \Re^*$.

Conversely, if

$$\mu(A) = \bar{\mu}(A),$$

then for arbitrary $\epsilon > 0$ there exist A', $A'' \in \Re$ such that

$$A' \subseteq A \subseteq A'',$$
$$\mu(A) - m(A') < \epsilon/2,$$
$$m(A'') - \bar{\mu}(A) < \epsilon/2,$$
$$m(A'' \smallsetminus A') = m(A'') - m(A') < \epsilon,$$

i.e., $A \in \Re^*$.

The following theorem holds for sets of $\mathfrak{M}$.

THEOREM 3. *If $A \subseteq \bigcup_{k=1}^{n} A_k$, then $\bar{\mu}(A) \leq \sum_{k=1}^{n} \bar{\mu}(A_k)$.*

Proof. Choose an A_k' such that

$$A_k \subseteq A_k', \qquad m(A_k') \leq \bar{\mu}(A_k) + \epsilon/2^k,$$

and let $A' = \bigcup_{k=1}^{n} A_k'$. Then

$$m(A') \leq \sum_{k=1}^{n} m(A_k') \leq \sum_{k=1}^{n} \bar{\mu}(A_k) + \epsilon,$$
$$\bar{\mu}(A) \leq \sum_{k=1}^{n} \bar{\mu}(A_k) + \epsilon;$$

since ϵ is arbitrary, $\bar{\mu}(A) \leq \sum_{k=1}^{n} \bar{\mu}(A_k)$.

THEOREM 4. *If $A_k \subseteq A$ $(1 \leq k \leq n)$ and $A_i \cap A_j = \emptyset$, then*

$$\mu(A) \geq \sum_{k=1}^{n} \mu(A_k).$$

Proof. Choose $A_k' \subseteq A_k$ such that $m(A_k') \geq \mu(A_k) - \epsilon/2^k$ and let $A' = \bigcup_{k=1}^{n} A_k'$. Then $A_i' \cap A_j' = \emptyset$ and

$$m(A') = \sum_{k=1}^{n} m(A_k') \geq \sum_{k} \mu(A_k) - \epsilon.$$

Since $A' \subseteq A$, $\mu(A) \geq m(A') \geq \sum_{k=1}^{n} \mu(A_k) - \epsilon$. Since $\epsilon > 0$ is arbitrary, $\mu(A) \geq \sum_{k=1}^{n} \mu(A_k)$.

We now define the function μ with domain of definition

$$S_\mu = \Re^*$$

as the common value of the inner and outer measures:

$$\mu(A) = \underline{\mu}(A) = \bar{\mu}(A).$$

Theorems 3 and 4 and the obvious fact that

$$\bar{\mu}(A) = \underline{\mu}(A) = m(A) \qquad\qquad (A \in \Re)$$

imply

THEOREM 5. *The function $\mu(A)$ is a measure and is an extension of the measure m.*

The construction we have discussed above is applicable to an arbitrary measure m defined on a ring.

The collection $S_{m_2} = \mathfrak{S}$ of elementary sets in the plane is essentially connected with the coordinate system: the sets of the collection $\mathfrak{S}$ consist of the rectangles whose sides are parallel to the coordinate axes. In the transition to the Jordan measure

$$J^{(2)} = j(m_2)$$

this dependence on the choice of the coordinate system vanishes: if $\{\bar{x}_1, \bar{x}_2\}$ is a system of coordinates related to the original coordinate system $\{x_1, x_2\}$ by the orthogonal transformation

$$\bar{x}_1 = x_1 \cos \alpha + x_2 \sin \alpha + a_1,$$

$$\bar{x}_2 = -x_1 \sin \alpha + x_2 \cos \alpha + a_2,$$

we obtain the same Jordan measure

$$J^{(2)} = j(m_2) = j(\bar{m}_2),$$

where $\bar{m}_2$ denotes the measure constructed by means of rectangles with sides parallel to the axes $\bar{x}_1, \bar{x}_2$. This fact is justified by the following general theorem:

THEOREM 6. *In order that two Jordan extensions $\mu_1 = j(m_1)$ and $\mu_2 = j(m_2)$ of measures m_1 and m_2 defined on rings $\Re_1$ and $\Re_2$ coincide, it is necessary and sufficient that*

$$\Re_1 \subseteq S_{\mu_2}, \qquad m_1(A) = \mu_2(A) \quad \text{on } \Re_1,$$

$$\Re_2 \subseteq S_{\mu_1}, \qquad m_2(A) = \mu_1(A) \quad \text{on } \Re_2.$$

The necessity is obvious. We shall prove the sufficiency.

Suppose that $A \in S_{\mu_1}$. Then there exist $A', A'' \in \Re_1$ such that

$$A' \subseteq A \subseteq A'', \qquad m_1(A'') - m_1(A') < \epsilon/3,$$

and $m_1(A') \leq \mu_1(A) \leq m_1(A'')$. By hypothesis, $\mu_2(A') = m_1(A')$ and $\mu_2(A'') = m_1(A'')$.

In view of the definition of the measure μ_2 there exist sets B', $B'' \in \Re_2$ such that

$$A' \supseteq B', \quad \mu_2(A') - m_2(B') < \epsilon/3;$$
$$B'' \supseteq A'', \quad m_2(B'') - \mu_2(A'') < \epsilon/3.$$

Hence

$$B' \subseteq A \subseteq B'',$$

and, obviously,

$$m_2(B'') - m_2(B') < \epsilon.$$

Since $\epsilon > 0$ is arbitrary, $A \in S_{\mu_2}$; and the relations

$$\mu_1(B') = m_2(B') \leq \mu_2(A) \leq m_2(B'') = \mu_1(B'')$$

imply that μ_2 is an extension of μ_1. Similarly, one shows that μ_1 is an extension of μ_2, and therefore

$$\mu_2(A) = \mu_1(A).$$

This proves the theorem.

Now, to show the independence of the Jordan measure in the plane of the choice of the coordinate system we need merely show that the set obtained from an elementary set by a rotation through an angle α is Jordan measurable. It is left to the reader to carry out this proof.

If the original measure m is defined on a semi-ring instead of a ring, it is natural to call the measure

$$j(m) = j(r(m))$$

obtained by extending m over the ring $\Re(S_m)$ and then extending the latter to a Jordan measure, the Jordan extension of m.

EXERCISES

1. If AB is a line segment in the plane, then $J^{(2)}(AB) = 0$.

2. Let ABC be a right triangle in the plane, with AB perpendicular to BC and with AB and BC parallel to the x- and y-axes, respectively.

a) Show that ABC is $J^{(2)}$-measurable.

b) Using the invariance of $J^{(2)}$ under translation and reflection in an axis, show that $J^{(2)}(ABC) = \frac{1}{2}(AB)(BC)$.

3. a) It follows from Ex. 2 and the text that any triangle is $J^{(2)}$-measurable and that its $J^{(2)}$-measure is the classical area.

b) Show, therefore, that a regular polygon is $J^{(2)}$-measurable and receives its classical area.

c) It follows now that a circle, i.e., a closed disk, is $J^{(2)}$-measurable.

4. Show that the plane set $A = \{(x, y): x^2 + y^2 \leq 1, x, y \text{ rational}\}$ is not $J^{(2)}$-measurable.

§37. Complete additivity. The general problem of the extension of measures

It is often necessary to consider countable unions as well as finite unions. Therefore, the condition of additivity we imposed on a measure (§34, Def. 1) is insufficient, and it is natural to replace it by the stronger condition of complete additivity.

DEFINITION 1. A measure μ is said to be *completely additive* (or *σ-additive*) if $A, A_1, \cdots, A_n, \cdots \in S_\mu$, where S_μ is the collection of sets on which μ is defined, and

$$A = \bigcup_{n=1}^{\infty} A_n, \qquad A_i \cap A_j = \emptyset \qquad\qquad (i \neq j)$$

imply that

$$\mu(A) = \sum_{n=1}^{\infty} \mu(A_n).$$

The plane Lebesgue measure constructed in §33 is σ-additive (Theorem 9). An altogether different example of a σ-additive measure may be constructed in the following way. Let

$$X = \{x_1, x_2, \cdots\}$$

be an arbitrary countable set and let the numbers $p_n > 0$ be such that

$$\sum_{n=1}^{\infty} p_n = 1.$$

The set S_μ consists of all the subsets of X. For each $A \subseteq X$ set

$$\mu(A) = \sum_{x_n \in A} p_n.$$

It is easy to verify that $\mu(A)$ is a σ-additive measure and that $\mu(X) = 1$. This example appears naturally in many problems in the theory of probability.

We shall also give an example of a measure which is additive, but not σ-additive. Let X be the set of all rational points on the closed interval $[0, 1]$ and let S_μ consist of the intersections of X with arbitrary intervals $(a, b), [a, b], [a, b)$ or $(a, b]$. It is easily seen that S_μ is a semi-ring. For each set $A_{ab} \in S_\mu$ set

$$\mu(A_{ab}) = b - a.$$

Then μ is an additive measure. It is not σ-additive since, for instance, $\mu(X) = 1$, but X is the union of a countable number of points each of which has measure zero.

In this and the succeeding two sections we shall consider σ-additive measures and their σ-additive extensions.

THEOREM 1. *If a measure m defined on a semi-ring S_m is completely additive, then its extension $\mu = r(m)$ to the ring $\Re(S_m)$ is completely additive.*

Proof. Suppose that

$$A \in \Re(S_m), \quad B_n \in \Re(S_m) \quad (n = 1, 2, \cdots)$$

and that

$$A = \bigcup_{n=1}^{\infty} B_n,$$

where $B_s \cap B_r = \emptyset$ ($s \neq r$). Then there exist sets A_j, $B_{ni} \in S_m$ such that

$$A = \bigcup_j A_j, \quad B_n = \bigcup_i B_{ni},$$

where the sets on the right-hand sides of these equalities are pairwise disjoint and the unions are finite unions (§34, Theorem 3).

Let $C_{nij} = B_{ni} \cap A_j$. It is easy to see that the sets C_{nij} are pairwise disjoint and that

$$A_j = \bigcup_n \bigcup_i C_{nij},$$

$$B_{ni} = \bigcup_j C_{nij}.$$

Therefore, because of the complete additivity of m on S_m,

(1) $$m(A_j) = \sum_n \sum_i m(C_{nij}),$$

(2) $$m(B_{ni}) = \sum_j m(C_{nij});$$

and because of the definition of $r(m)$ on $\Re(S_m)$,

(3) $$\mu(A) = \sum_j m(A_j),$$

(4) $$\mu(B_n) = \sum_i m(B_{ni}).$$

Relations (1), (2), (3) and (4) imply that $\mu(A) = \sum_n \mu(B_n)$. (The sums over i and j are finite sums, and the series over n converge.)

It could be proved that the Jordan extension of a σ-additive measure is σ-additive, but it is not necessary to do so because it will follow from the theory of Lebesgue extensions discussed in the next section.

THEOREM 2. *If a measure μ is σ-additive and $A, A_1, \cdots, A_n, \cdots \in S_\mu$, then*

$$A \subseteq \bigcup_{n=1}^{\infty} A_n$$

implies that

$$\mu(A) \leq \sum_{n=1}^{\infty} \mu(A_n).$$

Proof. Because of Theorem 1 it is sufficient to carry out the proof for a measure defined on a ring, since the validity of Theorem 2 for $\mu = r(m)$ immediately implies its applicability to the measure m. If S_μ is a ring, the sets

$$B_n = (A \cap A_n) \setminus \bigcup_{k=1}^{n-1} A_k$$

are elements of S_μ. Since

$$A = \bigcup_{n=1}^{\infty} B_n, \qquad B_n \subseteq A_n,$$

and the sets B_n are pairwise disjoint,

$$\mu(A) = \sum_{n=1}^{\infty} \mu(B_n) \leq \sum_{n=1}^{\infty} \mu(A_n).$$

In the sequel, we shall consider only σ-additive measures, without men tioning this fact explicitly.

$\star$ We have considered two ways of extending measures. In extending a measure m over the ring $\Re(S_m)$ in §35 we noted the uniqueness of the extension. The same is true for the Jordan extension $j(m)$ of an arbitrary measure m. If a set A is Jordan measurable with respect to a measure m (that is, $A \in S_{j(m)}$), then $\mu(A) = J(A)$, where μ is an arbitrary extension of m defined on A and $J = j(m)$ is the Jordan extension of m. It can be proved that an extension of m to a collection larger than $S_{j(m)}$ is not unique. More precisely, the following is true. Call a set A a *set of unicity* for a meas ure m if

1) there exists an extension of m defined on A;
2) for two such extensions μ_1 and μ_2,

$$\mu_1(A) = \mu_2(A).$$

Then the following theorem holds.

The set of sets of unicity of a measure m coincides with the collection of Jordan measurable sets relative to m, i.e., the collection of sets $S_{j(m)}$.

However, if we consider only σ-additive measures and their (σ-additive) extensions, then the collection of sets of unicity will, in general, be larger.

Since we shall be exclusively occupied with σ-additive measures in the sequel, we introduce

DEFINITION 2. A set A is said to be a *set of σ-unicity* for a σ-additive measure μ if

1) there exists a σ-additive extension λ of μ defined on A (that is, $A \in S_\lambda$);
2) if λ_1, λ_2 are two such σ-additive extensions, then

$$\lambda_1(A) = \lambda_2(A).$$

If A is a set of σ-unicity for a σ-additive measure μ, then the definition implies that if there is a σ-additive extension $\lambda(A)$ of μ defined on A, it is unique. $\star$

EXERCISES

1. Suppose μ is a completely additive measure on the collection of all subsets of a countable set X. Show that $\mu(A) = 0$ for all $A \subseteq X$ if, and

only if, $\mu(\{x\}) = 0$ for all $x \in X$, i.e., μ vanishes on every set consisting of a single point.

2. If X is a countable set, $\mathfrak{S}$ the class of all subsets of X and μ a completely additive measure defined on $\mathfrak{S}$, then μ must necessarily have the form of the second example after Def. 1 in §37, where, however, the p_n need only satisfy the conditions $p_n \geq 0$, $\sum p_n < \infty$.

3. Show that the measure defined in Ex. 2, §35 is completely additive. Hint: Imitate the procedure used in §33, Theorem 2.

4. Let $\mathfrak{S}$ be the semi-ring of left-closed right-open intervals on the line: $\mathfrak{S} = \{[a, b)\}$, let $F(t)$ be a nondecreasing left continuous real-valued function defined on the line and let μ_F be the Lebesgue measure defined in Remark 4 at the end of §33. Show that μ_F is completely additive by following a procedure analogous to that of Ex. 3 above. It will be necessary to show the following: Suppose that $a < b$. For $\epsilon > 0$ there exist c and d such that

$$a \leq c < d < b, \qquad [c, d] \subset [a, b)$$

and

$$\mu_F([c, d)) = F(d) - F(c) > F(b) - F(a) - \epsilon = \mu_F([a, b)) - \epsilon.$$

Similarly, there exist e, f such that $e < a < b \leq f$, $[a, b) \subset (e, f)$ and

$$\mu_F([e, f)) = F(f) - F(e) < F(b) - F(a) + \epsilon = \mu_F([a, b)) + \epsilon.$$

§38. The Lebesgue extension of a measure defined on a semi-ring with unity

Although the Jordan extension applies to a wide class of sets, it is nevertheless inadequate in many cases. Thus, for instance, if we take as the initial measure the area defined on the semi-ring of rectangles and consider the Jordan extension of this measure, then so comparatively simple a set as the set of points whose coordinates are rational and satisfy the condition $x^2 + y^2 \leq 1$ is not Jordan measurable.

The extension of a σ-additive measure defined on a semi-ring to a class of sets, which is maximal in a certain sense, can be effected by means of the Lebesgue extension. In this section we consider the Lebesgue extension of a measure defined on a semi-ring with a unit. The general case will be considered in §39.

The construction given below is to a considerable extent a repetition in abstract terms of the construction of the Lebesgue measure for plane sets in §33.

Let m be a σ-additive measure defined on a semi-ring S_m with a unit E. We define on the system $\mathfrak{S}$ of all subsets of E the functions $\mu^*(A)$ and $\mu_*(A)$ as follows.

DEFINITION 1. The *outer measure* of a set $A \subseteq E$ is

$$\mu^*(A) = \inf\{\textstyle\sum_n m(B_n); A \subseteq \bigcup_n B_n\},$$

where the lower bound is extended over all coverings of A by countable (finite or denumerable) collections of sets $B_n \in S_m$.

DEFINITION 2. The *inner measure* of a set $A \subseteq E$ is

$$\mu_*(A) = m(E) - \mu^*(E \setminus A).$$

Theorem 2 of §35 implies that $\mu_*(A) \leq \mu^*(A)$.

DEFINITION 3. A set $A \subseteq E$ is said to be (*Lebesgue*) *measurable* if

$$\mu_*(A) = \mu^*(A).$$

If A is measurable, the common value of $\mu_*(A) = \mu^*(A)$ is denoted by $\mu(A)$ and called the (*Lebesgue*) *measure* of A.

Obviously, if A is measurable, its complement $E \setminus A$ is also measurable. Theorem 2 of §37 immediately implies that

$$\mu_*(A) \leq \mu(A) \leq \mu^*(A)$$

for an arbitrary σ-additive extension μ of m. Therefore, for a Lebesgue measurable set A every σ-additive extension μ of m (if it exists) is equal to the common value of $\mu_*(A) = \mu^*(A)$. The Lebesgue measure is thus the σ-additive extension of m to the collection of all sets measurable in the sense of Def. 3. The definition of measurable set can obviously also be formulated as follows:

DEFINITION 3′. A set $A \subseteq E$ is said to be measurable if

$$\mu^*(A) + \mu^*(E \setminus A) = m(E).$$

It is expedient to use together with the initial measure m its extension $m' = r(m)$ (see §35) over the ring $\Re(S_m)$. It is clear that Def. 1 is equivalent to

DEFINITION 1′. The outer measure of a set A is

$$\mu^*(A) = \inf\{\textstyle\sum_n m'(B_n'); A \in \bigcup_n B_n'\} \qquad [B_n' \in \Re(S_m)].$$

In fact, since m' is σ-additive (§37, Theorem 1), an arbitrary sum $\sum_n m'(B_n')$, where $B_n' \in \Re(S_m)$, can be replaced by an equivalent sum

$$\textstyle\sum_{n,k} m(B_{nk}) \qquad (B_{nk} \in S_m),$$

where

$$B_n' = \bigcup_k B_{nk}, \qquad B_{ni} \cap B_{nj} = \emptyset \qquad (i \neq j).$$

The following is fundamental for the sequel.

THEOREM 1. *If*

$$A \subseteq \bigcup_n A_n,$$

where $\{A_n\}$ is a countable collection of sets then,

$$\mu^*(A) \le \sum_n \mu^*(A_n).$$

THEOREM 2. *If $A \in \Re$, then $\mu_*(A) = m'(A) = \mu^*(A)$, i.e., all the sets of $\Re(S_m)$ are measurable and their outer and inner measures coincide with m'.*

THEOREM 3. *A set A is measurable if, and only if, for arbitrary $\epsilon > 0$ there exists a set $B \in \Re(S_m)$ such that*

$$\mu^*(A \triangle B) < \epsilon.$$

These propositions were proved in §33 for plane Lebesgue measure (§33, Theorems 3–5). The proofs given there carry over verbatim to the general case considered here, and so we shall not repeat them.

THEOREM 4. *The collection $\mathfrak{M}$ of all measurable sets is a ring.*

Proof. Since

$$A_1 \cap A_2 = A_1 \setminus (A_1 \setminus A_2),$$

$$A_1 \cup A_2 = E \setminus [(E \setminus A_1) \cap (E \setminus A_2)],$$

t is sufficient to prove the following. If $A_1, A_2 \in \mathfrak{M}$, then

$$A = A_1 \setminus A_2 \in \mathfrak{M}.$$

Let A_1 and A_2 be measurable; then there exist $B_1, B_2 \in \Re(S_m)$ such that

$$\mu^*(A_1 \triangle B_1) < \epsilon/2, \qquad \mu^*(A_2 \triangle B_2) < \epsilon/2.$$

Setting $B = B_1 \setminus B_2 \in \Re(S_m)$ and using the relation

$$(A_1 \setminus A_2) \triangle (B_1 \setminus B_2) \subseteq (A_1 \triangle B_1) \cup (A_2 \triangle B_2),$$

we obtain

$$\mu^*(A \triangle B) < \epsilon.$$

Hence, A is measurable.

REMARK. It is obvious that E is the unit of the ring $\mathfrak{M}$, so that the latter is an algebra.

THEOREM 5. *The function $\mu(A)$ is additive on the set $\mathfrak{M}$ of measurable sets.*

The proof of this theorem is a verbatim repetition of the proof of Theorem 7, §33.

THEOREM 6. *The function $\mu(A)$ is σ-additive on the set $\mathfrak{M}$ of measurable sets.*

Proof. Let

$$A = \bigcup_{n=1}^{\infty} A_n \qquad (A, A_i \in \mathfrak{M}, \quad A_i \cap A_j = \emptyset \text{ if } i \ne j).$$

By Theorem 1,

(1) $$\mu^*(A) \leq \sum_n \mu(A_n),$$

and by Theorem 5, for arbitrary N,

$$\mu^*(A) \geq \mu^*(\textstyle\bigcup_{n=1}^N A_n) = \sum_{n=1}^N \mu^*(A_n).$$

Hence,

(2) $$\mu^*(A) \geq \sum_n \mu(A_n).$$

The theorem follows from (1) and (2).

We have therefore proved that the function $\mu(A)$ defined on $\mathfrak{M}$ possesses all the properties of a σ-additive measure.

This justifies the following

DEFINITION 4. The *Lebesgue extension* $\mu = L(m)$ *of a measure* $m(A)$ is the function $\mu(A)$ defined on the collection $S_\mu = \mathfrak{M}$ of measurable sets and coinciding on this collection with the outer measure $\mu^*(A)$.

In §33, in considering plane Lebesgue measure, we showed that not only finite, but denumerable unions and intersections of measurable sets are measurable. This is also true in the general case, that is,

THEOREM 7. *The collection* $\mathfrak{M}$ *of Lebesgue measurable sets is a Borel algebra with unit* E.

Proof. Since

$$\textstyle\bigcap_n A_n = E \setminus \bigcup_n (E \setminus A_n),$$

and since the complement of a measurable set is measurable, it is sufficient to prove the following: If $A_1, A_2, \cdots, A_n, \cdots \in \mathfrak{M}$, then $A = \bigcup_n A_n \in \mathfrak{M}$. The proof of this assertion is the same as that of Theorem 8, §33, for plane sets.

As in the case of plane Lebesgue measure, the σ-additivity of the measure implies its continuity, that is, if μ is a σ-additive measure defined on a B-algebra, and $A_1 \supseteq A_2 \supseteq \cdots \supseteq A_n \supseteq \cdots$ is a decreasing sequence of measurable sets, with

$$A = \textstyle\bigcap_n A_n,$$

then

$$\mu(A) = \lim_{n \to \infty} \mu(A_n);$$

and if $A_1 \subseteq A_2 \subseteq \cdots \subseteq A_n \subseteq \cdots$ is an increasing sequence of measurable sets, with

$$A = \textstyle\bigcup_n A_n,$$

then

$$\mu(A) = \lim_{n \to \infty} \mu(A_n).$$

The proof of this is the same as that of Theorem 10, §33, for plane measure.

⋆ 1. From the results of §37 and §38 we easily conclude that every Jordan measurable set A is Lebesgue measurable, and that its Jordan and Lebesgue measures are equal. It follows immediately that the Jordan extension of a σ-additive measure is σ-additive.

2. Every Lebesgue measurable set A is a set of unicity for the initial measure m. In fact, for arbitrary $\epsilon > 0$ there exists a set $B \in \Re$ such that $\mu^*(A \triangle B) < \epsilon$. For every extension λ of m defined on A,

$$\lambda(B) = m'(B),$$

since the extension of m on $\Re = \Re(S_m)$ is unique. Furthermore,

$$\lambda(A \triangle B) \leq \mu^*(A \triangle B) < \epsilon;$$

consequently,

$$| \lambda(A) - m'(B) | < \epsilon.$$

Hence,

$$| \lambda_1(A) - \lambda_2(A) | < 2\epsilon$$

for any two extensions $\lambda_1(A)$, $\lambda_2(A)$ of m. Therefore,

$$\lambda_1(A) = \lambda_2(A).$$

It can be proved that the class of Lebesgue measurable sets includes *all* the sets of unicity of an initial measure m.

3. Let m be a σ-additive measure defined on S, and let $\mathfrak{M} = L(S)$ be the domain of definition of its Lebesgue extension. It easily follows from Theorem 3 of this section that if S_1 is a semi-ring such that

$$S \subseteq S_1 \subseteq \mathfrak{M},$$

then

$$L(S_1) = L(S). \star$$

EXERCISES

1. Show that the collection of subsets A of E for which $\mu(A) = 0$ or $\mu(E \setminus A) = 0$ form a Borel algebra with E as a unit. This algebra is a subalgebra of $\mathfrak{M}$.

2. With the notation of the text for $A \subseteq E$ let

$$\mu_1^*(A) = \sup \{\mu^*(B) : B \subseteq A\}.$$

Show that $\mu_1^*(A) = \mu^*(A)$.

3. Suppose that $A \subseteq E$. Then $A \in \mathfrak{M}$ if, and only if, $A \cap B \in \mathfrak{M}$ for all $B \in S_m$.

4. For $A \subseteq E$, $A \in \mathfrak{M}$ if, and only if, for $\epsilon > 0$ there exist sets B_1, $B_2 \in \mathfrak{M}$ such that $B_1 \subset A \subset B_2$ and $\mu(B_2 \setminus B_1) < \epsilon$. [See §36, Def. 1 (Jordan measurability); also compare with §33, Ex. 6.]

5. For any $A \subseteq E$ we have
   a) $\mu^*(A) = \inf \{\mu B : A \subset B, \ B \in \mathfrak{M}\}$.
   b) $\mu_*(A) = \sup \{\mu B : B \subset A, \ B \in \mathfrak{M}\}$.

We see therefore that (abstract) Lebesgue measure is (abstractly) regular (see §33, Ex. 5).

6. Let S_{m_1}, S_{m_2} be two semi-rings on X with the same unit E; let m_1, m_2 be σ-additive measures defined on S_{m_1}, S_{m_2}, respectively; and let μ_1^*, μ_2^* be the outer measures on the set of all subsets of E defined by using m_1, m_2, respectively. Then $\mu_1^*(A) = \mu_2^*(A)$ for every $A \subseteq E$ if, and only if, $\mu_1^*(A) = m_2(A)$ for $A \in S_{m_2}$ and $\mu_2^*(A) = m_1(A)$ for $A \in S_{m_1}$. (See §36, Theorem 6 for the analogous theorem on Jordan extensions.)

§39. Extension of Lebesgue measures in the general case

If the semi-ring S_m on which the initial measure m is defined has no unit, the discussion of §38 must be modified. Def. 1 of the outer measure is retained, but the outer measure μ^* is now defined only on the collection $S_{\mu*}$ of the sets A for which there exists a covering $\bigcup_n B_n'$ of sets of S_m with a finite sum

$$\sum_n m(B_n).$$

Def. 2 becomes meaningless. The lower measure can also be defined (in several other ways) in the general case, but we shall not do so. It is more expedient at this point to define a measurable set in terms of the condition given in Theorem 3.

DEFINITION 1. A set A is said to be *measurable* if for arbitrary $\epsilon > 0$ there exists a set $B \in \mathfrak{R}(S_m)$ such that $\mu^*(A \triangle B) < \epsilon$.

Theorems 4, 5, 6 and Def. 4 of the preceding section remain true. The existence of a unit was used only in the proof of Theorem 4. To reprove Theorem 4 in the general case, it is necessary to show again that $A_1, A_2 \in \mathfrak{M}$ implies that $A_1 \cup A_2 \in \mathfrak{M}$. The proof of this is carried out in the same way as for $A_1 \setminus A_2$ on the basis of the relation.

$$(A_1 \cup A_2) \triangle (B_1 \cup B_2) \subseteq (A_1 \triangle B_1) \cup (A_2 \triangle B_2).$$

If S_m has no unit, Theorem 7 of §38 changes to

THEOREM 1. *For arbitrary initial measure m the collection $\mathfrak{M} = S_{L(m)}$ of Lebesgue measurable sets is a δ-ring; a set $A = \bigcup_{n=1}^{\infty} A_n$, where the sets A_n are measurable, is measurable if, and only if, the measure $\mu(\bigcup_{n=1}^{N} A_n)$ is bounded by a constant independent of N.*

The proof of this theorem is left to the reader.

REMARK. In our exposition the measure is always finite, so that the necessity of the last condition of the theorem is obvious.

Theorem 1 implies the

COROLLARY. The collection $\mathfrak{M}_A$ of all sets $B \in \mathfrak{M}$ which are subsets of a fixed set $A \in \mathfrak{M}$ is a Borel algebra. For instance, the collection of all Lebesgue measurable subsets (in the sense of the usual Lebesgue measure $\mu^{(1)}$ on the real line) of an arbitrary closed interval $[a, b]$ is a Borel algebra.

In conclusion we note yet another property of Lebesgue measures.

DEFINITION 2. A measure μ is said to be *complete* if $\mu(A) = 0$ and $A' \subseteq A$ imply that $A' \in S_\mu$.

It is clear that in that case $\mu(A') = 0$. It can be proved without difficulty that the Lebesgue extension of an arbitrary measure is complete. This follows from the fact that $A' \subseteq A$ and $\mu(A) = 0$ imply that $\mu^*(A') = 0$, and from the fact that an arbitrary set C for which $\mu^*(C) = 0$ is measurable, since $\emptyset \in R$ and

$$\mu^*(C \mathbin{\Delta} \emptyset) = \mu^*(C) = 0.$$

★ Let us indicate a connection between the method of constructing the Lebesgue extension of a measure and the method of completing a metric space. To this end, we note that $m'(A \mathbin{\Delta} B)$ can be thought of as the distance between the elements A, B of the ring $\mathfrak{R}(S_m)$. Then $\mathfrak{R}(S_m)$ becomes a metric space (in general, not complete) and its completion, according to Theorem 3 of §38, consists precisely of all the measurable sets. (In this connection, however, the sets A and B are not distinct, as points of a metric space, if $\mu(A \mathbin{\Delta} B) = 0$.) ★

EXERCISES

1. With the notation of the last paragraph of this section show that μ is a continuous function on the metric space $\mathfrak{M}$.

Chapter VI

MEASURABLE FUNCTIONS

§40. Definition and fundamental properties of measurable functions

Let X and Y be two sets and suppose that $\mathfrak{S}$ and $\mathfrak{S}'$ are classes of sub-sets of X, Y, respectively. An abstract function $y = f(x)$ defined on X, with values in Y, is said to be $(\mathfrak{S}, \mathfrak{S}')$-measurable if $A \in \mathfrak{S}'$ implies that $f^{-1}(A) \in \mathfrak{S}$.

For instance, if both X and Y are the real line D^1 (so that $f(x)$ is a real-valued function of a real variable), and $\mathfrak{S}$, $\mathfrak{S}'$ are the systems of all open (or closed) subsets of D^1, then the above definition of measurable function reduces to the definition of continuous function (see §12). If $\mathfrak{S}$ and $\mathfrak{S}'$ are the collections of all Borel sets, then the definition is that of B-measurable (Borel measurable) functions.

In the sequel our main interest in measurable functions will be from the point of view of integration. Fundamental to this point of view is the concept of the μ-measurability of real functions defined on a set X, with $\mathfrak{S}$ the collection of all μ-measurable subsets of X, and $\mathfrak{S}'$ the class of all B-sets on the real line. For simplicity, we shall assume that X is the unit of the domain of definition S_μ of the measure μ. Since, in view of the results of §38, every σ-additive measure can be extended to a Borel algebra, it is natural to assume that S_μ is a Borel algebra to begin with. Hence, for real functions we formulate the definition of measurability as follows:

DEFINITION 1. A real function $f(x)$ defined on a set X is said to be μ-measurable if

$$f^{-1}(A) \in S_\mu$$

for every Borel set A on the real line.

We denote by $\{x:Q\}$ the set of all $x \in X$ with property Q. We have the following

THEOREM 1. In order that a function $f(x)$ be μ-measurable it is necessary and sufficient that for every real c the set $\{x:f(x) < c\}$ be μ-measurable (that is, that this set be an element of S_μ).

Proof. The necessity of the condition is obvious, since the half-line $(-\infty, c)$ is a Borel set. To show the sufficiency we note first that the Borel closure $B(\Sigma)$ of the set of all the half-lines $(-\infty, c)$ coincides with the set B^1 of all Borel sets on the real line. By hypothesis, $f^{-1}(\Sigma) \subseteq S_\mu$. But then

$$f^{-1}(B(\Sigma)) = B(f^{-1}(\Sigma)) \subseteq B(S_\mu).$$

However, $B(S_\mu) = S_\mu$, since, by hypothesis, S_μ is a B-algebra. This proves the theorem.

THEOREM 2. *The pointwise limit function of a sequence of μ-measurable functions is μ-measurable.*

Proof. Suppose that $f_n(x) \to f(x)$. Then

(1) $$\{x : f(x) < c\} = \bigcup_k \bigcup_n \bigcap_{m>n} \{x : f_m(x) < c - 1/k\}.$$

For, if $f(x) < c$, there exists a k such that $f(x) < c - 2/k$; furthermore, for this k there is a sufficiently large n such that

$$f_m(x) < c - 1/k$$

for $m \geq n$. This means that x is an element of the set defined by the right-hand side of (1).

Conversely, if x is an element of the right-hand side of (1), then there exists a k such that

$$f_m(x) < c - 1/k$$

for all sufficiently large m. But then $f(x) < c$, that is, x belongs to the set on the left-hand side of (1).

If the functions $f_n(x)$ are measurable, the sets

$$\{x : f_m(x) < c - 1/k\}$$

are elements of S_μ. Since S_μ is a Borel algebra, the set

$$\{x : f(x) < c\}$$

also belongs to S_μ in virtue of (1). This proves that $f(x)$ is measurable.

For the further discussion of measurable functions it is convenient to represent each such function as the limit of a sequence of simple functions.

DEFINITION 2. A function $f(x)$ is said to be *simple* if it is μ-measurable and if it assumes no more than a countable set of values.

It is clear that the concept of simple function depends on the choice of the measure μ.

The structure of simple functions is characterized by the following theorem:

THEOREM 3. *A function $f(x)$ which assumes no more than a countable set of distinct values*

$$y_1, \cdots, y_n, \cdots$$

is μ-measurable if, and only if, all the sets

$$A_n = \{x : f(x) = y_n\}$$

are μ-measurable.

Proof. The necessity of the condition is clear, since each A_n is the inverse image of a set consisting of one point y_n, and every such set is a Borel set. The sufficiency follows from the fact that, by hypothesis, the inverse image $f^{-1}(B)$ of an arbitrary set $B \subseteq D^1$ is the union $\bigcup_{y_n \in B} A_n$ of no more than a countable number of measurable sets A_n, that is, it is measurable.

The further use of simple functions will be based on the following theorem:

THEOREM 4. *In order that a function $f(x)$ be μ-measurable it is necessary and sufficient that it be representable as the limit of a uniformly convergent sequence of simple functions.*

Proof. The sufficiency is clear from Theorem 2. To prove the necessity we consider an arbitrary measurable function $f(x)$ and set $f_n(x) = m/n$, where $m/n \leq f(x) < (m + 1)/n$ (m an integer, n a natural number). It is clear that the functions $f_n(x)$ are simple; they converge uniformly to $f(x)$ as $n \to \infty$, since $|f(x) - f_n(x)| \leq 1/n$.

THEOREM 5. *The sum of two μ-measurable functions is μ-measurable.*

Proof. We prove the theorem first for simple functions. If $f(x)$ and $g(x)$ are two simple functions assuming the values

$$f_1, \cdots, f_n, \cdots,$$

$$g_1, \cdots, g_n, \cdots,$$

respectively, then their sum $f(x) + g(x)$ can assume only the values $h = f_i + g_j$, where each of these values is assumed on the set

$$(2) \qquad \{x : h(x) = h\} = \bigcup_{f_i + g_j = h}(\{x : f(x) = f_i\} \cap \{x : g(x) = g_j\}).$$

The possible number of values of h is countable, and the corresponding set $\{x : h(x) = h\}$ is measurable, since the right side of (2) is obviously a measurable set.

To prove the theorem for arbitrary measurable functions $f(x)$ and $g(x)$ we consider sequences of simple functions $\{f_n(x)\}$ and $\{g_n(x)\}$ converging to $f(x)$ and $g(x)$, respectively. Then the simple functions $f_n(x) + g_n(x)$ converge uniformly to the function $f(x) + g(x)$. The latter, in view of Theorem 4, is measurable.

THEOREM 6. *A B-measurable function of a μ-measurable function is μ-measurable.*

Proof. Let $f(x) = \varphi[\psi(x)]$, where φ is Borel measurable and ψ is μ-measurable. If $A \subseteq D^1$ is an arbitrary B-measurable set, then its inverse image $A' = \varphi^{-1}(A)$ is B-measurable, and the inverse image $A'' = \psi^{-1}(A')$ of A' is μ-measurable. Since $f^{-1}(A) = A''$, it follows that f is measurable.

Theorem 6 is applicable, in particular, to continuous functions φ (they are always B-measurable).

THEOREM 7. *The product of μ-measurable functions is μ-measurable.*

Proof. Since $fg = \frac{1}{4}[(f + g)^2 - (f - g)^2]$, the theorem follows from Theorems 5 and 6 and the fact that $\varphi(t) = t^2$ is continuous.

EXERCISE. Show that if $f(x)$ is measurable and nonvanishing, then $1/f(x)$ is also measurable.

In the study of measurable functions it is often possible to neglect the values of the function on a set of measure zero. In this connection, we introduce the following

DEFINITION. Two functions f and g defined on the same measurable set E are said to be *equivalent* (notation: $f \sim g$) if

$$\mu\{x : f(x) \neq g(x)\} = 0.$$

We say that a property is satisfied *almost everywhere* (abbreviated a.e.) on E if it is satisfied at all points of E except for a set of measure zero. Hence, two functions are equivalent if they are equal a.e.

THEOREM 8. *If two functions f and g, continuous on a closed interval E, are equivalent, they are equal.*

Proof. Let us suppose that $f(x_0) \neq g(x_0)$, i.e., $f(x_0) - g(x_0) \neq 0$. Since $f - g$ is continuous, $f - g$ does not vanish in some neighborhood of x_0. This neighborhood has positive measure; hence

$$\mu\{x : f(x) \neq g(x)\} > 0,$$

that is, the continuous functions f and g cannot be equivalent if they differ even at a single point.

Obviously, the equivalence of two arbitrary measurable (that is, in general, discontinuous) functions does not imply their equality; for instance, the function equal to 1 at the rational points and 0 at the irrational points is equivalent to the function identically zero on the real line.

THEOREM 9. *A function $f(x)$ defined on a measurable set E and equivalent on E to a measurable function $g(x)$ is measurable.*

In fact, it follows from the definition of equivalence that the sets

$$\{x : f(x) > a\}, \quad \{x : g(x) > a\}$$

may differ only on a set of measure zero; consequently, if the second set is measurable, so is the first.

★ The above definition of a measurable function is quite formal. In 1913 Luzin proved the following theorem, which shows that a measurable function is a function which in a certain sense can be approximated by a continuous function.

LUZIN'S THEOREM. *In order that a function $f(x)$ be measurable on a closed interval $[a, b]$ it is necessary and sufficient that for every $\epsilon > 0$ there exist a* *function $\varphi(x)$ continuous on $[a, b]$ such that*

$$\mu\{x\!:\!f(x) \neq \varphi(x)\} \leq \epsilon.$$

In other words, a measurable function can be made into a continuous function by changing its values on a set of arbitrarily small measure. This property, called by Luzin the C-property, may be taken as the definition of a measurable function.★

EXERCISES

1. For $A \subseteq X$ let χ_A be the characteristic function of A defined by $\chi_A(x) = 1$ if $x \in A$, $\chi_A(x) = 0$ if $x \in X \setminus A$.

a) $\chi_{A \cap B}(x) = \chi_A(x)\chi_B(x),$

$\chi_{A \cup B}(x) = \chi_A(x) + \chi_B(x) - \chi_A(x)\chi_B(x),$

$\chi_{A \triangle B}(x) = |\chi_A(x) - \chi_B(x)|,$

$\chi_{\emptyset}(x) \equiv 0, \qquad \chi_X(x) \equiv 1,$

$\chi_A(x) \leq \chi_B(x) \ (x \in X)$ if, and only if, $A \subseteq B$.

b) $\chi_A(x)$ is μ-measurable if, and only if, $A \in S_\mu$.

2. Suppose $f(x)$ is a real-valued function of a real variable. If $f(x)$ is nondecreasing, then $f(x)$ is Borel measurable.

3. Let $X = [a, b]$ be a closed interval on the real line. If $f(x)$ is defined on X and $X = \bigcup_{i=1}^{n} E_i$, where each E_i is a subinterval of X, $E_i \cap E_j = \emptyset$ and $f(E_i) = y_i$, then we call f a *step function*.

Suppose that f is nondecreasing (or nonincreasing) on X. Show that all the functions of the approximating sequence of simple functions $\{f_n\}$ of Theorem 4 of this section are step functions.

4. Assume that $X = [a, b]$ contains a non-Lebesgue measurable set A. Define a function $f(x)$ on X such that $|f(x)|$ is Lebesgue measurable, but $f(x)$ is not.

5. Two real functions $f(x)$ and $g(x)$ defined on a set X are both μ-measurable. Show that $\{x\!:\!f(x) = g(x)\}$ is μ-measurable.

6. Let X be a set containing two or more points. Suppose that $\mathfrak{S} = \{\emptyset, X\}$. Describe all measurable functions.

7. Let $f(x)$ be a μ-measurable function defined on X. For t real define $\varphi(t) = \mu(\{x\!:\!f(x) < t\})$. Show that φ is monotone nondecreasing, continuous on the left, $\lim_{t \to -\infty} \varphi(t) = 0$, and $\lim_{t \to \infty} \varphi(t) = \mu(X)$. φ is called the *distribution function* of $f(x)$.

§41. Sequences of measurable functions. Various types of convergence

Theorems 5 and 7 of the preceding section show that the arithmetical operations applied to measurable functions again yield measurable functions. According to Theorem 2 of §40, the class of measurable functions, unlike the class of continuous functions, is also closed under passage to a limit. In addition to the usual pointwise convergence, it is expedient to define certain other types of convergence for measurable functions. In this section we shall consider these definitions of convergence, their basic properties and the relations between them.

DEFINITION 1. A sequence of functions $f_n(x)$ defined on a measure space X (that is, a space with a measure defined in it) is said to *converge to a function $F(x)$ a.e.* if

(1) $$\lim_{n\to\infty} f_n(x) = F(x)$$

for almost all $x \in X$ [that is, the set of x for which (1) does not hold is of measure zero].

EXAMPLE. The sequence of functions $f_n(x) = (-x)^n$ converges to the function $F(x) \equiv 0$ a.e. on the closed interval [0, 1] (indeed, everywhere except at the point $x = 1$).

Theorem 2 of §40 admits of the following generalization.

THEOREM 1. *If a sequence $\{f_n(x)\}$ of μ-measurable functions converges to a function $F(x)$ a.e., then $F(x)$ is measurable.*

Proof. Let A be the set on which

$$\lim_{n\to\infty} f_n(x) = F(x).$$

By hypothesis, $\mu(E \setminus A) = 0$. The function $F(x)$ is measurable on A by Theorem 2 of §40. Since every function is obviously measurable on a set of measure zero, $F(x)$ is measurable on $(E \setminus A)$; consequently, it is measurable on E.

EXERCISE. Suppose that a sequence of measurable functions $f_n(x)$ converges a.e. to a limit function $f(x)$. Prove that the sequence $f_n(x)$ converges a.e. to $g(x)$ if, and only if, $g(x)$ is equivalent to $f(x)$.

The following theorem, known as Egorov's theorem, relates the notions of convergence a.e. and uniform convergence.

THEOREM 2. *Suppose that a sequence of measurable functions $f_n(x)$ converges to $f(x)$ a.e. on E. Then for every $\delta > 0$ there exists a measurable set $E_\delta \subset E$ such that*

1) $\mu(E_\delta) > \mu(E) - \delta$;
2) *the sequence $f_n(x)$ converges to $f(x)$ uniformly on E_δ.*

Proof. According to Theorem 1, $f(x)$ is measurable. Set

$$E_n{}^m = \bigcap_{i \geq n} \{x : |f_i(x) - f(x)| < 1/m\}.$$

Hence, $E_n{}^m$ for fixed m and n is the set of all x for which

$$|f_i(x) - f(x)| < 1/m \qquad (i \geq n).$$

Let

$$E^m = \bigcup_n E_n{}^m.$$

It is clear from the definition of the sets $E_n{}^m$ that

$$E_1{}^m \subseteq E_2{}^m \subseteq \cdots \subseteq E_n{}^m \subseteq \cdots$$

for fixed m. Therefore, since a σ-additive measure is continuous (see §38), for arbitrary m and $\delta > 0$ there exists an $n(m)$ such that

$$\mu(E^m \setminus E_{n(m)}{}^m) < \delta/2^m.$$

We set

$$E_\delta = \bigcap_m E_{n(m)}{}^m$$

and prove that E_δ is the required set.

We shall prove first that the sequence $\{f_i(x)\}$ converges uniformly to $f(x)$ on E_δ. This follows at once from the fact that if $x \in E_\delta$, then

$$|f_i(x) - f(x)| < 1/m \qquad\qquad (i \geq n(m))$$

for arbitrary m. We now estimate the measure of the set $E \setminus E_\delta$. To do so we note that $\mu(E \setminus E^m) = 0$ for every m. In fact, if $x_0 \in E \setminus E^m$, then

$$|f_i(x_0) - f(x_0)| \geq 1/m$$

for infinitely many values of i, that is, the sequence $\{f_n(x)\}$ does not converge to $f(x)$ at $x = x_0$. Since $\{f_n(x)\}$ converges to $f(x)$ a.e. by hypothesis,

$$\mu(E \setminus E^m) = 0.$$

Hence,

$$\mu(E \setminus E_{n(m)}{}^m) = \mu(E^m \setminus E_{n(m)}{}^m) < \delta/2^m.$$

Therefore,

$$\begin{aligned}
\mu(E \setminus E_\delta) &= \mu(E \setminus \bigcap_m E_{n(m)}{}^m) \\
&= \mu(\bigcup_m (E \setminus E_{n(m)}{}^m)) \\
&\leq \sum_m \mu(E \setminus E_{n(m)}{}^m) \\
&< \sum_{m=1}^{\infty} \delta/2^m = \delta.
\end{aligned}$$

This proves the theorem.

★ DEFINITION 2. A sequence of measurable functions $f_n(x)$ converges in measure to a function $F(x)$ if for every $\sigma > 0$

$$\lim_{n \to \infty} \mu\{x : |f_n(x) - F(x)| \geq \sigma\} = 0.$$

Theorems 3 and 4 below relate the concepts of convergence a.e. and convergence in measure.

THEOREM 3. *If a sequence of measurable functions $f_n(x)$ converges a.e. to a function $F(x)$, then it converges in measure to $F(x)$.*

Proof. Theorem 1 implies that the limit function $F(x)$ is measurable. Let A be the set (of measure zero) on which $f_n(x)$ does not converge to $F(x)$. Furthermore, let

$$E_k(\sigma) = \{x : |f_k(x) - F(x)| \geq \sigma\}, \qquad R_n(\sigma) = \bigcup_{k=n}^{\infty} E_k(\sigma),$$
$$M = \bigcap_{n=1}^{\infty} R_n(\sigma).$$

It is clear that all these sets are measurable. Since

$$R_1(\sigma) \supset R_2(\sigma) \supset \cdots,$$

and because of the continuity of the measure,

$$\mu(R_n(\sigma)) \to \mu(M) \qquad\qquad (n \to \infty)$$

We now verify that

(2) $$M \subseteq A.$$

In fact, if $x_0 \notin A$, that is, if

$$\lim_{n\to\infty} f_n(x_0) = F(x_0),$$

then for every $\sigma > 0$ there is an n such that

$$|f_n(x_0) - F(x_0)| < \sigma,$$

that i $x_0 \notin E_n(\sigma)$; hence $x_0 \notin M$.

But since $\mu(A) = 0$, it follows from (2) that $\mu(M) = 0$. Consequently,

$$\mu(R_n(\sigma)) \to 0 \qquad\qquad (n \to \infty).$$

Since $E_n(\sigma) \subseteq R_n(\sigma)$, this proves the theorem.

It is easy to see by an example that convergence in measure does not imply convergence a.e. For each natural number k define k functions

$$f_1^{(k)}, \cdots, f_k^{(k)}$$

on the half-open interval $(0, 1]$ as follows:

$$f_i^{(k)}(x) = \begin{cases} 1 & (i-1)/k < x \leq i/k, \\ 0 & \text{for the remaining values of } x. \end{cases}$$

Writing these functions in a sequence yields a sequence which, as is easily verified, converges in measure to zero, but converges nowhere (prove this!).

EXERCISE. Suppose that a sequence of measurable functions $f_n(x)$ converges in measure to a limit function $f(x)$. Prove that the sequence $f_n(x)$ converges in measure to $g(x)$ if, and only if, $g(x)$ is equivalent to $f(x)$.

Although the above example shows that the full converse of Theorem 3 is not true, nevertheless we have the following

THEOREM 4. *Suppose that a sequence of measurable functions $f_n(x)$ converges in measure to $f(x)$. Then the sequence $\{f_n(x)\}$ contains a subsequence $\{f_{n_k}(x)\}$ which converges a.e. to $f(x)$.*

Proof. Let $\epsilon_1, \cdots, \epsilon_n, \cdots$ be a sequence of positive numbers such that

$$\lim_{n \to \infty} \epsilon_n = 0,$$

and suppose that the positive numbers $\eta_1, \cdots, \eta_n, \cdots$ are such that the series

$$\eta_1 + \eta_2 + \cdots$$

converges. We construct a sequence of indices

$$n_1 < n_2 < \cdots$$

as follows: n_1 is a natural number such that

$$\mu\{x: |f_{n_1}(x) - f(x)| \geq \epsilon_1\} < \eta_1$$

(such an n_1 necessarily exists). Then n_2 is chosen so that

$$\mu\{x: |f_{n_2}(x) - f(x)| \geq \epsilon_2\} < \eta_2 \qquad (n_2 > n_1).$$

In general, n_k is a natural number such that

$$\mu\{x: |f_{n_k}(x) - f(x)| \geq \epsilon_k\} < \eta_k \qquad (n_k > n_{k-1}).$$

We shall show that the subsequence $\{f_{n_k}(x)\}$ converges to $f(x)$ a.e. In fact, let

$$R_i = \bigcup_{k=i}^{\infty} \{x: |f_{n_k}(x) - f(x)| \geq \epsilon_k\}, \qquad Q = \bigcap_{i=1}^{\infty} R_i.$$

Since

$$R_1 \supset R_2 \supset R_3 \supset \cdots \supset R_n \supset \cdots,$$

and the measure is continuous, it follows that $\mu(R_i) \to \mu(Q)$.

On the other hand, it is clear that $\mu(R_i) < \sum_{k=i}^{\infty} \eta_k$, whence $\mu(R_i) \to 0$ as $i \to \infty$. Since $\mu(R_i) \to 0$,

$$\mu(Q) = 0.$$

It remains to verify that

$$f_{n_k}(x) \to f(x)$$

for all $x \in E \setminus Q$. Suppose that $x_0 \in E \setminus Q$. Then there is an i_0 such that $x_0 \notin R_{i_0}$. Then

$$x_0 \notin \{x: |f_{n_k}(x) - f(x)| \geq \epsilon_k\}$$

for all $k \geq i_0$, i.e.,

$$|f_{n_k}(x_0) - f(x_0)| < \epsilon_k .$$

Since $\epsilon_k \to 0$ by hypothesis,

$$\lim_{k \to \infty} f_{n_k}(x_0) = f(x_0).$$

This proves the theorem.★

EXERCISES

1. Egorov's theorem does not yield the result that there exists a subset $E_0 \subseteq E$ with $\mu E_0 = 0$ and that the sequence $\{f_n(x)\}$ converges uniformly to $f(x)$ on $E \setminus E_0$. However, prove that there exists a sequence $\{E_i\}$ of measurable subsets of E such that $\mu(E \setminus \bigcup_i E_i) = 0$ and on each E_i the convergence is uniform.

2. Suppose that $\{f_n\}$, f are measurable functions defined on E and for $\epsilon > 0$ there exists a measurable set $F \subseteq E$ such that $\mu F < \epsilon$ and $\{f_n(x)\}$ converges uniformly to $f(x)$ in $E \setminus F$. Show that $\{f_n(x)\}$ converges to $f(x)$ a.e. in E.

3. Let X be the set of positive integers, $\mathfrak{S}$ the class of all subsets of X, and for $A \in \mathfrak{S}$ let $\mu(A)$ be the number of points in A. Note that we are here allowing sets of infinite measure. If χ_n is the characteristic function of $\{1, \cdots, n\}$, then $\chi_n(x)$ converges everywhere to $\chi_X(x) \equiv 1$, but the conclusion of Egorov's theorem does not hold.

4. A sequence of measurable functions $f_n(x)$ is said to be *fundamental in measure* if for every $\sigma > 0$,

$$\lim_{m,n \to \infty} \mu\{x : |f_n(x) - f_m(x)| > \sigma\} = 0.$$

Show that if $\{f_n(x)\}$ is fundamental in measure, then there exists a measurable function $f(x)$ such that $\{f_n(x)\}$ converges in measure to $f(x)$. Hint: Use Theorem 4.

5. Let $\{A_n\}$ be a sequence of measurable sets and let χ_n be the characteristic function of A_n. Show that the sequence $\{\chi_n\}$ is fundamental in measure if, and only if, $\lim_{m,n \to \infty} \mu(A_n \triangle A_m) = 0$.

6. If $\{f_n(x)\}$, $\{g_n(x)\}$ converge in measure to $f(x)$ and $g(x)$, respectively, then $\{f_n(x) + g_n(x)\}$ converges in measure to $f(x) + g(x)$.

Chapter VII

THE LEBESGUE INTEGRAL

In the preceding chapter we considered the fundamental properties of measurable functions, which are a very broad generalization of continuous functions. The classical definition of the integral, the Riemann integral, is, in general, not applicable to the class of measurable functions. For instance, the well known Dirichlet function (equal to zero at the irrational points and one at the rational points) is obviously measurable, but not Riemann integrable. Therefore, the Riemann integral is not suitable for measurable functions.

The reason for this is perfectly clear. For simplicity, let us consider functions on a closed interval. To define the Riemann integral we divide the interval on which a function $f(x)$ is defined into small subintervals and, choosing a point ξ_k in each of these subintervals, form the sum

$$\sum_k f(\xi_k) \Delta x_k .$$

What we do, essentially, is to replace the value of $f(x)$ at each point of the closed interval $\Delta x_k = [x_k , x_{k+1}]$ by its value at an arbitrarily chosen point ξ_k of this interval. But this, of course, can be done only if the values of $f(x)$ at points which are close together are also close together, i.e., if $f(x)$ is continuous or if its set of discontinuities is "not too large." (A bounded function is Riemann integrable if, and only if, its set of discontinuities has measure zero.)

The basic idea of the Lebesgue integral, in contrast to the Riemann integral, is to group the points x not according to their nearness to each other on the x-axis, but according to the nearness of the values of the function at these points. This at once makes it possible to extend the notion of integral to a very general class of functions.

In addition, a single definition of the Lebesgue integral serves for functions defined on arbitrary measure spaces, while the Riemann integral is introduced first for functions of one variable, and is then generalized, with appropriate changes, to the case of several variables.

In the sequel, without explicit mention, we consider a σ-additive measure $\mu(A)$ defined on a Borel algebra with unit X. The sets $A \subseteq X$ of the algebra are μ-measurable, and the functions $f(x)$—defined for all $x \in X$—are also μ-measurable.

§42. The Lebesgue integral of simple functions

We introduce the Lebesgue integral first for the simple functions, that is, for measurable functions whose set of values is countable.

Let $f(x)$ be a simple function with values

$$y_1, \cdots, y_n, \cdots \qquad (y_i \neq y_j \text{ for } i \neq j).$$

It is natural to define the integral of $f(x)$ over (on) a set A as

(1) $$\int_A f(x) \, d\mu = \sum_n y_n \mu \{x : x \in A, f(x) = y_n\}.$$

We therefore arrive at the following definition.

DEFINITION. A simple function $f(x)$ is μ-*integrable* over A if the series (1) is absolutely convergent. If $f(x)$ is integrable, the sum of the series (1) is called the *integral* of $f(x)$ over A.

In this definition it is assumed that all the y_n are distinct. However, it is possible to represent the value of the integral of a simple function as a sum of products $c_k \mu(B_k)$ without assuming that all the c_k are distinct. This can be done by means of the

LEMMA. *Suppose that* $A = \bigcup_k B_k$, $B_i \cap B_j = \emptyset$ $(i \neq j)$ *and that* $f(x)$ *assumes only one value on each set* B_k. *Then*

(2) $$\int_A f(x) \, d\mu = \sum_k c_k \mu(B_k),$$

where the function $f(x)$ *is integrable over* A *if, and only if, the series* (2) *is absolutely convergent.*

Proof. It is easy to see that each set

$$A_n = \{x : x \in A, f(x) = y_n\}$$

is the union of all the sets B_k for which $c_k = y_n$. Therefore,

$$\sum_n y_n \mu(A_n) = \sum_n y_n \sum_{c_k = y_n} \mu(B_k) = \sum_k c_k \mu(B_k).$$

Since the measure is nonnegative,

$$\sum_n |y_n| \mu(A_n) = \sum_n |y_n| \sum_{c_k = y_n} \mu(B_k) = \sum_k |c_k| \mu(B_k),$$

that is, the series $\sum y_n \mu(A_n)$ and $\sum_k c_k \mu(B_k)$ are either both absolutely convergent or both divergent.

We shall now derive some properties of the Lebesgue integral of simple functions.

A) $$\int_A f(x) \, d\mu + \int_A g(x) \, d\mu = \int_A \{f(x) + g(x)\} \, d\mu,$$

where the existence of the integrals on the left side implies the existence of the integral on the right side.

To prove A) we assume that $f(x)$ assumes the values f_i on the sets

$F_i \subseteq A$, and that $g(x)$ assumes the values g_j on the sets $G_j \subseteq A$; hence

(3) $$J_1 = \int_A f(x) \, d\mu = \sum_i f_i \mu(F_i),$$

(4) $$J_2 = \int_A g(x) \, d\mu = \sum_j g_j \mu(G_j).$$

Then, by the lemma,

(5) $$J = \int_A \{f(x) + g(x)\} \, d\mu = \sum_i \sum_j (f_i + g_j) \mu(F_i \cap G_j).$$

But

$$\mu(F_i) = \sum_j \mu(F_i \cap G_j),$$
$$\mu(G_j) = \sum_i \mu(F_i \cap G_j),$$

so that the absolute convergence of the series (3) and (4) implies the absolute convergence of the series (5). Hence

$$J = J_1 + J_2.$$

B) For every constant k,

$$k \int_A f(x) \, d\mu = \int_A \{kf(x)\} \, d\mu,$$

where the existence of the integral on the left implies the existence of the integral on the right. (The proof is immediate.)

C) A simple function $f(x)$ bounded on a set A is integrable over A, and

$$\left| \int_A f(x) \, d\mu \right| \leq M \mu(A),$$

where $|f(x)| \leq M$ on A. (The proof is immediate.)

EXERCISES

1. If A, B are measurable subsets of X, then

$$\int_X |\chi_A(x) - \chi_B(x)| \, d\mu = \mu(A \, \Delta \, B).$$

2. If the simple function $f(x)$ is integrable over A and $B \subseteq A$, then $f(x)$ is integrable over B.

3. Let $F_0 = [0, 1]$. Define the simple function $f(x)$ on F_0 as follows: On the 2^{n-1} open intervals deleted in the nth stage of the construction of the Cantor set F let $f(x) = n$. On F let $f(x) = 0$. Compute $\int_{F_0} f(x) \, d\mu$, where μ is linear Lebesgue measure.

§43. The general definition and fundamental properties of the Lebesgue integral

DEFINITION. We shall say that a function $f(x)$ is *integrable* over a set A if there exists a sequence of simple functions $f_n(x)$ integrable over A and uniformly convergent to $f(x)$. The limit

$$(1) \qquad J = \lim_{n \to \infty} \int_A f_n(x) \, d\mu$$

is denoted by

$$\int_A f(x) \, d\mu$$

and is called the integral of $f(x)$ over A.

This definition is correct if the following conditions are satisfied:

1. The limit (1) for an arbitrary uniformly convergent sequence of simple functions integrable over A exists.

2. This limit, for fixed $f(x)$, is independent of the choice of the sequence $\{f_n(x)\}$.

3. For simple functions this definition of integrability and of the integral is equivalent to that of §42.

All these conditions are indeed satisfied.

To prove the first it is enough to note that because of Properties A), B) and C) of integrals of simple functions,

$$\left| \int_A f_n(x) \, d\mu - \int_A f_m(x) \, d\mu \right| \leq \mu(A) \sup \{|f_n(x) - f_m(x)| ; x \in A\}.$$

To prove the second condition it is necessary to consider two sequences $\{f_n(x)\}$ and $\{f_n^*(x)\}$ and to use the fact that

$$\left| \int_A f_n(x) \, d\mu - \int_A f_n^*(x) \, d\mu \right|$$

$$\leq \mu(A)\{\sup [|f_n(x) - f(x)| ; x \in A] + \sup [|f_n^*(x) - f(x)| ; x \in A]\}.$$

Finally, to prove the third condition it is sufficient to consider the sequence $f_n(x) = f(x)$.

We shall derive the fundamental properties of the Lebesgue integral.

THEOREM 1.

$$\int_A 1 \cdot d\mu = \mu(A).$$

Proof. This is an immediate consequence of the definition.

THEOREM 2. *For every constant k,*

$$k \int_A f(x) \, d\mu = \int_A \{kf(x)\} \, d\mu,$$

where the existence of the integral on the left implies the existence of the integral on the right.

Proof. To prove this take the limit in Property B) for simple functions.

THEOREM 3.

$$\int_A f(x) \, d\mu + \int_A g(x) \, d\mu = \int_A \{f(x) + g(x)\} \, d\mu,$$

where the existence of the integrals on the left implies the existence of the integral on the right.

The proof is obtained by passing to the limit in Property A) of integrals of simple functions.

THEOREM 4. *A function $f(x)$ bounded on a set A is integrable over A.*

The proof is carried out by passing to the limit in Property C).

THEOREM 5. *If $f(x) \geq 0$, then*

$$\int_A f(x) \, d\mu \geq 0,$$

on the assumption that the integral exists.

Proof. For simple functions the theorem follows immediately from the definition of the integral. In the general case, the proof is based on the possibility of approximating a nonnegative function by simple functions (in the way indicated in the proof of Theorem 4, §40).

COROLLARY 1. *If $f(x) \geq g(x)$, then*

$$\int_A f(x) \, d\mu \geq \int_A g(x) \, d\mu.$$

COROLLARY 2. *If $m \leq f(x) \leq M$ on A, then*

$$m\mu(A) \leq \int_A f(x) \, d\mu \leq M\mu(A).$$

THEOREM 6. *If*

$$A = \bigcup_n A_n \qquad (A_i \cap A_j = \emptyset \text{ for } i \neq j),$$

then

$$\int_A f(x) \, d\mu = \sum_n \int_{A_n} f(x) \, d\mu,$$

where the existence of the integral on the left implies the existence of the integrals and the absolute convergence of the series on the right.

Proof. We first verify the theorem for a simple function $f(x)$ which assumes the values

$$y_1, \cdots, y_k, \cdots.$$

Let

$$B_k = \{x : x \in A, f(x) = y_k\},$$
$$B_{nk} = \{x : x \in A_n, f(x) = y_k\}.$$

Then

$$\int_A f(x) \, d\mu = \sum_k y_k \mu(B_k) = \sum_k y_k \sum_n \mu(B_{nk})$$

(1)

$$= \sum_n \sum_k y_k \mu(B_{nk}) = \sum_n \int_{A_n} f(x) \, d\mu.$$

Since the series $\sum_k y_k \mu(B_k)$ is absolutely convergent if $f(x)$ is integrable, and the measures are nonnegative, all the other series in (1) also converge absolutely.

If $f(x)$ is an arbitrary function, its integrability over A implies that for every $\epsilon > 0$ there exists a simple function $g(x)$ integrable over A such that

(2) $$|f(x) - g(x)| < \epsilon.$$

For $g(x)$,

(3) $$\int_A g(x) \, d\mu = \sum_n \int_{A_n} g(x) \, d\mu,$$

where $g(x)$ is integrable over each of the sets A_n, and the series in (3) is absolutely convergent. The latter and the estimate (2) imply that $f(x)$ is also integrable over each A_n, and

$$\sum_n \left| \int_{A_n} f(x) \, d\mu - \int_{A_n} g(x) \, d\mu \right| \leq \sum_n \epsilon\mu(A_n) \leq \epsilon\mu(A),$$

$$\left| \int_A f(x) \, d\mu - \int_A g(x) \, d\mu \right| \leq \epsilon\mu(A).$$

This together with (3) yields the absolute convergence of the series

$$\sum_n \int_{A_n} f(x) \, d\mu$$

and the estimate

$$\left| \sum_n \int_{A_n} f(x) \, d\mu - \int_A f(x) \, d\mu \right| \leq 2\epsilon\mu(A).$$

Since $\epsilon > 0$ is arbitrary,

$$\sum_n \int_{A_n} f(x) \, d\mu = \int_A f(x) \, d\mu.$$

COROLLARY. *If $f(x)$ is integrable over A, then $f(x)$ is integrable over an arbitrary $A' \subseteq A$.*

THEOREM 7. *If a function $\varphi(x)$ is integrable over A, and $|f(x)| \leq \varphi(x)$, then $f(x)$ is also integrable over A.*

Proof. If $f(x)$ and $\varphi(x)$ are simple functions, then A can be written as the union of a countable number of sets on each of which $f(x)$ and $\varphi(x)$ are constant:

$$f(x) = a_n, \qquad \varphi(x) = \alpha_n \qquad (|a_n| \leq \alpha_n).$$

The integrability of $\varphi(x)$ implies that

$$\sum_n |a_n| \mu(A_n) \leq \sum_n \alpha_n \mu(A_n) = \int_A \varphi(x) \, d\mu.$$

Therefore $f(x)$ is also integrable, and

$$\left| \int_A f(x) \, d\mu \right| = |\sum_n a_n \mu(A_n)| \leq \sum_n |a_n| \mu(A_n)$$

$$= \int_A |f(x)| \, d\mu \leq \int_A \varphi(x) \, d\mu.$$

Passage to the limit proves the theorem in the general case.

TRANS. NOTE. The proof is as follows: For $\epsilon > 0$, choose an $n_0 > 1/\epsilon$. Let $\{\varphi_n : n \geq n_0\}$ be a sequence of integrable simple functions converging uniformly to the function $\varphi(x) + \epsilon$, and let $\{f_n : n \geq n_0\}$ be a sequence of simple functions converging uniformly to $f(x)$. These sequences are chosen so that they satisfy the inequalities

$$\varphi_n(x) \geq 0, \qquad |\varphi_n(x) - [\varphi(x) + \epsilon]| < 1/n, \qquad |f_n(x) - f(x)| < 1/n.$$

Then $|f_n(x)| < \varphi_n(x)$ and

$$\left| \int_A f_n(x) \, d\mu \right| \leq \int_A |f_n(x)| \, d\mu \leq \int_A \varphi_n(x) \, d\mu.$$

Since

$$\int_A \varphi_n(x) \, d\mu \to \int_A \varphi(x) \, d\mu + \epsilon \mu(A),$$

each $f_n(x)$ is integrable and $f(x)$ is integrable, and

$$\left| \int_A f(x) \, d\mu \right| \leq \int_A \varphi(x) \, d\mu + \epsilon \mu(A).$$

Since $\epsilon > 0$ is arbitrary, the desired inequality follows.

THEOREM 8. *The integrals*

$$J_1 = \int_A f(x)\,d\mu, \qquad J_2 = \int_A |f(x)|\,d\mu$$

either both exist or both do not exist.

Proof. The existence of J_2 implies the existence of J_1 by Theorem 7.

For a simple function the converse follows from the definition of the integral. The general case is proved by passing to the limit and noting that

$$||a| - |b|| \le |a - b|.$$

THEOREM 9 (THE CHEBYSHEV INEQUALITY). *If $\varphi(x) \ge 0$ on A, then*

$$\mu\{x : x \in A, \ \varphi(x) \ge c\} \le (1/c) \int_A \varphi(x)\,d\mu.$$

Proof. Setting

$$A' = \{x : x \in A, \varphi(x) \ge c\},$$

we have

$$\int_A \varphi(x)\,d\mu = \int_{A'} \varphi(x)\,d\mu + \int_{A \setminus A'} \varphi(x)\,d\mu \ge \int_{A'} \varphi(x)\,d\mu \ge c\mu(A').$$

COROLLARY. *If*

$$\int_A |f(x)|\,d\mu = 0,$$

then $f(x) = 0$ a.e.

For, by the Chebyshev inequality,

$$\mu\{x : x \in A, |f(x)| \ge 1/n\} \le n \int_A |f(x)|\,d\mu = 0$$

for all n. Therefore,

$$\mu\{x : x \in A, f(x) \ne 0\} \le \sum_{n=1}^{\infty} \mu\{x : x \in A, |f(x)| \ge 1/n\} = 0.$$

EXERCISES

1. Suppose $f(x)$ is integrable over E, and that F is a measurable subset of E. Then $\chi_F f$ is integrable over E and

$$\int_E \chi_F(x)f(x)\,d\mu = \int_F f(x)\,d\mu.$$

2 (FIRST MEAN VALUE THEOREM). Let $f(x)$ be measurable,

$$m \le f(x) \le M$$

on A, and suppose that $g(x) \geq 0$ is integrable over A. Then there exists a real number a such that $m \leq a \leq M$ and $\int_A f(x)g(x) \, d\mu = a \int_A g(x) \, d\mu$.

3. Suppose that $f(x)$ is integrable over the set $E = [a, b]$ and that μ is linear Lebesgue measure. Then $F(x) = \int_{[a,x]} f(x) \, d\mu$ is defined for $a \leq x \leq b$.

a) Show that

$$[F(x_2) - F(x_1)]/(x_2 - x_1) = [1/(x_2 - x_1)] \int_{[x_1,x_2]} f(x) \, d\mu$$

for $a \leq x_1 < x_2 \leq b$.

b) For any point x_0, $a < x_0 < b$, at which $f(x)$ is continuous show that $F'(x_0) = f(x_0)$.

4. Let f, g be integrable over E.

a) If $\int_A f(x) \, d\mu = \int_A g(x) \, d\mu$ for every measurable $A \subseteq E$, then $f(x) = g(x)$ a.e. on E.

b) If $\int_A f(x) \, d\mu = 0$, for every measurable $A \subseteq E$, then $f(x) = 0$ a.e. on E.

5. Suppose $E = [a, b]$, μ is Lebesgue measure and f is integrable over E. Show that $\int_{[a,c]} f(x) \, d\mu = 0$ for $a \leq c \leq b$ implies that $f(x) = 0$ a.e. on E. Hint: Consider the class $\mathfrak{S}$ of $A \subseteq E$ for which $\int_A f(x) \, d\mu = 0$ and apply the preceding exercise.

§44. Passage to the limit under the Lebesgue integral

The question of taking the limit under the integral sign, or, equivalently, the possibility of termwise integration of a convergent series often arises in various problems.

It is proved in classical analysis that a sufficient condition for interchanging limits in this fashion is the uniform convergence of the sequence (or series) involved.

In this section we shall prove a far-reaching generalization of the corresponding theorem of classical analysis.

THEOREM 1. *If a sequence $f_n(x)$ converges to $f(x)$ on A and*

$$|f_n(x)| \leq \varphi(x)$$

for all n, where $\varphi(x)$ is integrable over A, then the limit function $f(x)$ is in-

tegrable over A and

$$\int_A f_n(x)\, d\mu \to \int_A f(x)\, d\mu.$$

Proof. It easily follows from the conditions of the theorem that

$$|f(x)| \le \varphi(x).$$

Let $A_k = \{x : k - 1 \le \varphi(x) < k\}$, and let $B_m = \bigcup_{k \ge m+1} A_k = \{x : \varphi(x) \ge m\}$.

By Theorem 6 of §43,

$$(*) \qquad \int_A \varphi(x)\, d\mu = \sum_k \int_{A_k} \varphi(x)\, d\mu,$$

and the series $(*)$ converges absolutely.

Hence

$$\int_{B_m} \varphi(x)\, d\mu = \sum_{k \ge m+1} \int_{A_k} \varphi(x)\, d\mu.$$

The convergence of the series $(*)$ implies that there exists an m such that

$$\int_{B_m} \varphi(x)\, d\mu < \epsilon/5.$$

The inequality $\varphi(x) < m$ holds on $A \setminus B_m$. By Egorov's theorem, $A \setminus B_m$ can be written as $A \setminus B_m = C \cup D$, where $\mu(D) < \epsilon/5m$ and the sequence $\{f_n\}$ converges uniformly to f on C.

Choose an N such that

$$|f_n(x) - f(x)| < \epsilon/5\mu(C)$$

for all $n > N$ and $x \in C$. Then

$$\int_A [f_n(x) - f(x)]\, d\mu = \int_{B_m} f_n(x)\, d\mu - \int_{B_m} f(x)\, d\mu$$

$$+ \int_D f_n(x)\, d\mu - \int_D f(x)\, d\mu + \int_C [f_n(x) - f(x)]\, d\mu < 5\epsilon/5 = \epsilon.$$

COROLLARY. *If* $|f_n(x)| \le M$ *and* $f_n(x) \to f(x)$, *then*

$$\int_A f_n(x)\, d\mu \to \int_A f(x)\, d\mu.$$

REMARK. Inasmuch as the values assumed by a function on a set of measure zero do not affect the value of the integral, it is sufficient to assume in Theorem 1 that $\{f_n(x)\}$ converges to $f(x)$ a.e.

THEOREM 2. *Suppose that*

$$f_1(x) \leq f_2(x) \leq \cdots \leq f_n(x) \leq \cdots$$

on a set A, where the functions $f_n(x)$ are integrable and their integrals are bounded from above:

$$\int_A f_n(x) \, d\mu \leq K.$$

Then

(1) $$f(x) = \lim_{n \to \infty} f_n(x)$$

exists a.e. on A, $f(x)$ is integrable on A and

$$\int_A f_n(x) \, d\mu \to \int_A f(x) \, d\mu.$$

Clearly, the theorem also holds for a monotone descending sequence of integrable functions whose integrals are bounded from below.

On the set on which the limit (1) does not exist, $f(x)$ can be defined arbitrarily; for instance, we may set $f(x) = 0$ on this set.

Proof. We assume that $f(x) \geq 0$, since the general case is easily reduced to this case by writing

$$\bar{f}_n(x) = f_n(x) - f_1(x).$$

We consider the set

$$\Omega = \{x : x \in A, f_n(x) \to \infty\}.$$

It is easy to see that $\Omega = \bigcap_r \bigcup_n \Omega_n^{(r)}$, where

$$\Omega_n^{(r)} = \{x : x \in A, f_n(x) > r\}.$$

By the Chebyshev inequality (Theorem 9, §43),

$$\mu(\Omega_n^{(r)}) \leq K/r.$$

Since $\Omega_1^{(r)} \subseteq \Omega_2^{(r)} \subseteq \cdots \subseteq \Omega_n^{(r)} \subseteq \cdots$, it follows that

$$\mu(\bigcup_n \Omega_n^{(r)}) \leq K/r.$$

Further, since

$$\Omega \subseteq \bigcup_n \Omega_n^{(r)}$$

for every r, $\mu(\Omega) \leq K/r$. Since r is arbitrary,

$$\mu(\Omega) = 0.$$

This also proves that the monotone sequence $f_n(x)$ has a finite limit $f(x)$ a.e. on A.

Now let $\varphi(x) = r$ for all x such that

$$r - 1 \leq f(x) < r \qquad (r = 1, 2, \cdots).$$

If we prove that $\varphi(x)$ is integrable on A, the theorem will follow immediately from Theorem 1.

We denote by A_r the set of all points $x \in A$ for which $\varphi(x) = r$ and set

$$B_s = \bigcup_{r=1}^{s} A_r.$$

Since the functions $f_n(x)$ and $f(x)$ are bounded on B_s and $\varphi(x) \leq f(x) + 1$, it follows that

$$\int_{B_s} \varphi(x) \, d\mu \leq \int_{B_s} f(x) \, d\mu + \mu(A)$$

$$= \lim_{n \to \infty} \int_{B_s} f_n(x) \, d\mu + \mu(A) \leq K + \mu(A).$$

But

$$\int_{B_s} \varphi(x) \, d\mu = \sum_{r=1}^{s} r\mu(A_r).$$

Since the partial sums in the above equation are bounded, the series

$$\sum_{r=1}^{\infty} r\mu(A_r) = \int_A \varphi(x) \, d\mu$$

converges. Hence $\varphi(x)$ is integrable on A.

COROLLARY. *If $\psi_n(x) \geq 0$ and*

$$\sum_{n=1}^{\infty} \int_A \psi_n(x) \, d\mu < \infty,$$

then the series $\sum_{n=1}^{\infty} \psi_n(x)$ converges a.e. on A and

$$\int_A \left(\sum_{n=1}^{\infty} \psi_n(x) \right) d\mu = \sum_{n=1}^{\infty} \int_A \psi_n(x) \, d\mu.$$

THEOREM 3 (FATOU). *If a sequence of measurable nonnegative functions $\{f_n(x)\}$ converges a.e. on A to $f(x)$ and*

$$\int_A f_n(x) \, d\mu \leq K,$$

then $f(x)$ is integrable on A and

$$\int_A f(x) \, d\mu \leq K.$$

Proof. Set

$$\varphi_n(x) = \inf \{f_k(x); k \geq n\}.$$

$\varphi_n(x)$ is measurable, since

$$\{x : \varphi_n(x) < c\} = \mathsf{U}_{k \geq n} \{x : f_k(x) < c\}.$$

Furthermore, $0 \leq \varphi_n(x) \leq f_n(x)$, so that $\varphi_n(x)$ is integrable, and

$$\int_A \varphi_n(x) \, d\mu \leq \int_A f_n(x) \, d\mu \leq K.$$

Finally,

$$\varphi_1(x) \leq \varphi_2(x) \leq \cdots \leq \varphi_n(x) \leq \cdots$$

and

$$\lim_{n \to \infty} \varphi_n(x) = f(x) \qquad \text{(a.e.).}$$

The required result follows by application of the preceding theorem to $\{\varphi_n(x)\}$.

THEOREM 4. *If $A = \mathsf{U}_n A_n$, $A_i \cap A_j = \emptyset$ $(i \neq j)$ and the series*

$$(2) \qquad \sum_n \int_{A_n} |f(x)| \, d\mu$$

converges, then $f(x)$ is integrable on A and

$$\int_A f(x) \, d\mu = \sum_n \int_{A_n} f(x) \, d\mu.$$

What is new here as compared with Theorem 6, §43 is the assertion that the convergence of the series (2) implies the integrability of $f(x)$ on A.

We first prove that the theorem is true for a simple function $f(x)$, which assumes the values f_i on the sets B_i. Setting

$$A_{ni} = A_n \cap B_i,$$

we have

$$\int_{A_n} |f(x)| \, d\mu = \sum_i |f_i| \, \mu(A_{ni}).$$

The convergence of the series (2) implies that the series

$$\sum_n \sum_i |f_i| \, \mu(A_{ni}) = \sum_i |f_i| \, \mu(B_i \cap A)$$

converge.

In view of the convergence of the last series, the integral

$$\int_A f(x) \, d\mu = \sum_i f_i \mu(B_i \cap A)$$

exists.

In the general case, we approximate $f(x)$ by a simple function $\bar{f}(x)$ so that

(3) $$|f(x) - \bar{f}(x)| < \epsilon.$$

Then

$$\int_{A_n} |\bar{f}(x)| \, d\mu \leq \int_{A_n} |f(x)| \, d\mu + \epsilon\mu(A_n).$$

Since the series

$$\sum_n \mu(A_n) = \mu(A)$$

converges, the convergence of (2) implies the convergence of

$$\sum_n \int_{A_n} |\bar{f}(x)| \, d\mu,$$

that is, in view of what has just been proved, the integrability of the simple function $\bar{f}(x)$ on A. But then, by (3), $f(x)$ is also integrable on A.

EXERCISES

1. Let $X = [0, 1]$, let μ be linear Lebesgue measure and suppose that $f(x) \geq 0$ is measurable.

a) Suppose $\epsilon_1 > \epsilon_2 > \cdots > \epsilon_n > \cdots$, $\epsilon_n \to 0$ and $f(x)$ integrable over $[\epsilon_n, 1]$. Then f is integrable over $[0, 1]$ if, and only if, $\lim_{n\to\infty} \int_{[\epsilon_n,1]} f(x) \, d\mu$ exists, and in that case

$$\int_{[0,1]} f(x) \, d\mu = \lim_{n\to\infty} \int_{[\epsilon_n,1]} f(x) \, d\mu.$$

(This justifies the remark made at the end of §45.)

2. For $f(x)$ measurable on the measurable set E define

$$S_n = \sum_{k=-\infty}^{\infty} k2^{-n}\mu\{x: k2^{-n} \leq f(x) < (k+1)2^{-n}, x \in E\}, \quad n = 1, 2, \cdots.$$

a) If f is integrable on E, then each S_n is absolutely convergent, $\lim_n S_n$ exists and

$$\int_E f(x) \, d\mu = \lim_{n\to\infty} S_n.$$

b) Conversely, if S_n converges absolutely for some n, then S_n converges absolutely for all n, $f(x)$ is integrable over E and the above equality holds.

c) Let $n = 0$ in a). It follows that

$$S_0 = \sum_k |k| \, \mu\{x: k \leq f(x) < k+1\} < \infty.$$

Show consequently that f integrable over E implies that

$$\lim_{m\to\infty} m\mu\{x: |f(x)| \geq m, x \in E\} = 0.$$

Hint: Reduce the problem to the case $f(x) \geq 0$.

3. Let X be measurable, and let (a, b) be an interval of real numbers. Suppose $f(x, t)$ is real-valued for $x \in X$, $t \in (a, b)$ and that it satisfies the following:

(i) For $t \in (a, b)$, $f(x, t)$ is integrable over X.

(ii) $\partial f(x, t)/\partial t$ exists for all $t \in (a, b)$ and there exists a function $S(x)$ integrable over X for which

$$|\partial f(x, t)/\partial t| \leq S(x) \qquad [x \in X, t \in (a, b)].$$

Show that

$$d/dt \int_X f(x, t) \, d\mu = \int_X \partial f(x, t)/\partial t \, d\mu.$$

Hint: For $t_0 \in (a, b)$ the limit defining the derivative can be obtained by using a sequence $\{t_n\}$ in (a, b), $t_n \to t_0$. Apply Theorem 1.

§45. Comparison of the Lebesgue and Riemann integrals

We shall discuss the relation of the Lebesgue integral to the usual Riemann integral. In doing so, we restrict ourselves to the simplest case, linear Lebesgue measure on the real line.

THEOREM. *If the Riemann integral*

$$J = (R) \int_a^b f(x) \, dx$$

exists, then $f(x)$ *is Lebesgue integrable on* $[a, b]$, *and*

$$\int_{[a,b]} f(x) \, d\mu = J.$$

Proof. Consider the partition of $[a, b]$ into 2^n subintervals by the points

$$x_k = a + (k/2^n)(b - a)$$

and the Darboux sums

$$\bar{S}_n = (b - a)2^{-n} \sum_{k=1}^{2^n} M_{nk},$$

$$\underline{S}_n = (b - a)2^{-n} \sum_{k=1}^{2^n} m_{nk},$$

where M_{nk} is the least upper bound of $f(x)$ on the interval

$$x_{k-1} \leq x \leq x_k,$$

and m_{nk} is the greatest lower bound of $f(x)$ on the same interval. By definition, the Riemann integral is

$$J = \lim_{n\to\infty} \bar{S}_n = \lim_{n\to\infty} \underline{S}_n .$$

We set

$$\bar{f}_n(x) = M_{nk} \qquad\qquad (x_{k-1} \leq x < x_k),$$
$$\underline{f}_n(x) = m_{nk} \qquad\qquad (x_{k-1} \leq x < x_k).$$

The functions $\bar{f}_n$ and $\underline{f}_n$ can be extended to the point $x = b$ arbitrarily. It is easily verified that

$$\int_{[a,b]} \bar{f}_n(x)\ d\mu = \bar{S}_n,$$

$$\int_{[a,b]} \underline{f}_n(x)\ d\mu = \underline{S}_n .$$

Since $\{\bar{f}_n\}$ is a nonincreasing sequence and $\{\underline{f}_n\}$ is a nondecreasing sequence,

$$\bar{f}_n(x) \to \bar{f}(x) \geq f(x),$$
$$\underline{f}_n(x) \to \underline{f}(x) \leq f(x)$$

a.e. By Theorem 2 of §44,

$$\int_{[a,b]} \bar{f}(x)\ d\mu = \lim_{n\to\infty} \bar{S}_n = J = \lim_{n\to\infty} \underline{S}_n = \int_{[a,b]} \underline{f}(x)\ d\mu.$$

Therefore,

$$\int_{[a,b]} |\bar{f}(x) - \underline{f}(x)|\ d\mu = \int_{[a,b]} \{\bar{f}(x) - \underline{f}(x)\}\ d\mu = 0;$$

consequently,

$$\bar{f}(x) - \underline{f}(x) = 0$$

a.e., i.e.,

$$\bar{f}(x) = \underline{f}(x) = f(x),$$

$$\int_{[a,b]} f(x)\ d\mu = \int_{[a,b]} \bar{f}(x)\ d\mu = J.$$

This proves the theorem.

TRANS. NOTE. The following well known characterization of Riemann integrable functions now follows immediately from the preceding theorem and the observation that $f(x)$ is continuous at x if, and only if, $\bar{f}(x) = \underline{f}(x)$:

Let $f(x)$ be bounded on $[a, b]$. Then $f(x)$ is Riemann integrable if, and only if, it is continuous a.e.

It is easy to construct an example of a bounded function which is Lebesgue integrable but not Riemann integrable (for instance, the Dirichlet function mentioned above).

An arbitrary function $f(x)$ for which the Riemann integral

$$\int_{\epsilon}^{1} |f(x)| \, dx$$

approaches a finite limit J as $\epsilon \to 0$ is Lebesgue integrable on $[0, 1]$, and

$$\int_{[0,1]} f(x) \, d\mu = \lim_{\epsilon \to 0} \int_{\epsilon}^{1} f(x) \, dx.$$

(See Ex. 4, §44.)

In this connection it is interesting to note that the improper integrals

$$\int_{0}^{1} f(x) \, dx = \lim_{\epsilon \to 0} \int_{\epsilon}^{1} f(x) \, dx,$$

where

$$\lim_{\epsilon \to 0} \int_{\epsilon}^{1} |f(x)| \, dx = \infty,$$

cannot be taken in the sense of Lebesgue: Lebesgue integration is absolute integration in the sense of Theorem 8, §43.

EXERCISES

1. Let $f(x)$, $g(x)$ be Riemann integrable functions on $[a, b]$. Then $f(x)g(x)$ is Riemann integrable on $[a, b]$. This fact can be proved without the characterization of Riemann integrable functions given in the text, but a direct proof is difficult.

2. A nondecreasing (nonincreasing) real-valued function defined on an interval $[a, b]$ is Riemann integrable on this interval.

3. Show that the function

$$f(x) = d/dx(x^2 \sin 1/x^2) = 2x \sin 1/x^2 - (2/x) \cos 1/x^2$$

is not Lebesgue integrable over $[0, 1]$, although $f(x)$ is continuous on $[\epsilon, 1]$ for every $\epsilon > 0$ and $\lim_{\epsilon \to 0} \int_{\epsilon}^{1} f(x) \, dx$ exists; that is, $f(x)$ is improperly Riemann integrable, the integral being only conditionally convergent. Hint: $|f(x)| \geq (2/x) |\cos 1/x^2| - 2x \geq x^{-1} - 2x$ on each of the intervals $\{(2n + \frac{1}{4})\pi\}^{-\frac{1}{2}} \leq x \leq \{(2n - \frac{1}{4})\pi\}^{-\frac{1}{2}}$.

§46. Products of sets and measures

Theorems on the reduction of double (or multiple) integrals to repeated integrals play an important part in analysis. The fundamental result in the theory of multiple Lebesgue integrals is Fubini's theorem, which we shall prove in §48. We first introduce some auxiliary concepts and results which, however, have an interest independent of Fubini's theorem.

The set Z of ordered pairs (x, y), where $x \in X$, $y \in Y$, is called the *product* of the sets X and Y and is denoted by $X \times Y$. In the same way, the set Z of finite ordered sequences $(x_1, \cdots, x_n)$, where $x_k \in X_k$, is called the product of the sets $X_1, \cdots, X_n$ and is denoted by

$$Z = X_1 \times X_2 \times \cdots \times X_n = \times_{k=1}^{n} X_k.$$

In particular, if

$$X_1 = X_2 = \cdots = X_n = X,$$

the set Z is the nth power of the set X:

$$Z = X^n.$$

For instance, the n-dimensional coordinate space D^n is the nth power of the real line D^1. The unit cube J^n, that is, the set of points of D^n with co-ordinates satisfying the conditions

$$0 \leq x_k \leq 1 \qquad (1 \leq k \leq n),$$

is the nth power of the closed unit interval $J^1 = [0, 1]$.

If $\mathfrak{S}_1, \cdots, \mathfrak{S}_n$ are collections of subsets of the sets $X_1, \cdots, X_n$, then

$$\mathfrak{R} = \mathfrak{S}_1 \times \cdots \times \mathfrak{S}_n$$

is the collection of subsets of the set $X = \times_k X_k$ representable in the form

$$A = A_1 \times \cdots \times A_n \qquad (A_k \in \mathfrak{S}_k).$$

If $\mathfrak{S}_1 = \mathfrak{S}_2 = \cdots = \mathfrak{S}_n = \mathfrak{S}$, then $\mathfrak{R}$ is the nth power of $\mathfrak{S}$:

$$\mathfrak{R} = \mathfrak{S}^n.$$

For instance, the set of all parallelopipeds in D^n is the nth power of the set of closed intervals in D^1.

THEOREM 1. *If $\mathfrak{S}_1, \cdots, \mathfrak{S}_n$ are semi-rings, then $\mathfrak{R} = \times_k \mathfrak{S}_k$ is a semi-ring.*

Proof. In view of the definition of a semi-ring (§34), we must prove that if $A, B \in \mathfrak{R}$, then $A \cap B \in \mathfrak{R}$; and if, moreover, $B \subseteq A$, then $A = \bigcup_{i=1}^{m} C_i$, where $C_1 = B$, $C_i \cap C_j = \emptyset$ $(i \neq j)$ and $C_i \in \mathfrak{R}$ $(1 \leq i \leq m)$.

We shall carry out the proof for $n = 2$.

I) Suppose that $A, B \in \mathfrak{S}_1 \times \mathfrak{S}_2$. Then

$$A = A_1 \times A_2, \qquad A_1 \in \mathfrak{S}_1, \qquad A_2 \in \mathfrak{S}_2;$$
$$B = B_1 \times B_2, \qquad B_1 \in \mathfrak{S}_1, \qquad B_2 \in \mathfrak{S}_2.$$

Hence

$$A \cap B = (A_1 \cap B_1) \times (A_2 \cap B_2),$$

and since

$$A_1 \cap B_1 \in \mathfrak{S}_1, \qquad A_1 \cap B_2 \in \mathfrak{S}_2,$$

it follows that

$$A \cap B \in \mathfrak{S}_1 \times \mathfrak{S}_2.$$

II) Now, on the same assumptions as in I), suppose that $B \subseteq A$. Then

$$B_1 \subseteq A_1, \qquad B_2 \subseteq A_2,$$

and because $\mathfrak{S}_1$ and $\mathfrak{S}_2$ are semi-rings, it follows that

$$A_1 = B_1 \cup B^{(1)} \cup \cdots \cup B_1^{(k)},$$
$$A_2 = B_2 \cup B_2^{(1)} \cup \cdots \cup B_2^{(l)},$$
$$A = A_1 \times A_2 = (B_1 \times B_2) \cup (B_1 \times B_2^{(1)}) \cup \cdots \cup (B_1 \times B_2^{(l)})$$
$$\cup (B_1^{(1)} \times B_2) \cup (B_1^{(1)} \times B_2^{(1)}) \cup \cdots \cup (B_1^{(1)} \times B_2^{(l)})$$
$$\cdots\cdots\cdots\cdots\cdots\cdots\cdots\cdots\cdots\cdots\cdots\cdots\cdots\cdots\cdots\cdots\cdots$$
$$\cup (B_1^{(k)} \times B_2) \cup (B_1^{(k)} \times B_2^{(1)}) \cup \cdots \cup (B_1^{(k)} \times B_2^{(l)}).$$

In the last relation the first term is $B_1 \times B_2 = B$ and all the other terms are elements of $\mathfrak{S}_1 \times \mathfrak{S}_2$ (all pairwise disjoint). This proves the theorem.

However, if the $\mathfrak{S}_k$ are rings or Borel rings, it does not follow that $\times_k \mathfrak{S}_k$ is a ring or a Borel ring.

Suppose that the measures

$$\mu_1(A_1), \mu_2(A_2), \cdots, \mu_n(A_n) \qquad\qquad (A_k \in \mathfrak{S}_k)$$

are defined on the semi-rings $\mathfrak{S}_1, \cdots, \mathfrak{S}_n$.

We define the measure

$$\mu = \mu_1 \times \mu_2 \times \cdots \times \mu_n$$

on

$$\mathfrak{R} = \mathfrak{S}_1 \times \mathfrak{S}_2 \times \cdots \times \mathfrak{S}_n$$

by the following condition: If $A = A_1 \times \cdots \times A_n$, then

$$\mu(A) = \mu_1(A_1)\mu_2(A_2) \cdots \mu_n(A_n).$$

We must prove that $\mu(A)$ is a measure, that is, that $\mu(A)$ is additive. We do so for $n = 2$. Let

$$A = A_1 \times A_2 = \bigcup B^{(k)}, \qquad B^{(i)} \cap B^{(j)} = \emptyset \qquad (i \neq j),$$
$$B^{(k)} = B_1^{(k)} \times B_2^{(k)}.$$

It was shown in §34 that there are partitions

$$A_1 = \bigcup_m C_1^{(m)}, \qquad A_2 = \bigcup_n C_2^{(n)}$$

such that $B_1^{(k)} = \bigcup_{m \in M^{(k)}} C_1^{(m)}$ and $B_2^{(k)} = \bigcup_{n \in N^{(k)}} C_2^{(n)}$. It is obvious that

(1) $\mu(A) = \mu_1(A_1)\mu_2(A_2) = \sum_m \sum_n \mu_1(C_1^{(m)})\mu_2(C_2^{(n)}),$

(2) $\mu(B^{(k)}) = \mu_1(B_1^{(k)})\mu_2(B_2^{(k)}) = \sum_{m \in M^{(k)}} \sum_{n \in N^{(k)}} \mu_1(C_1^{(m)})\mu_2(C_2^{(n)}),$

where the right side of (1) contains just once all the terms appearing on the right side of (2). Therefore,

$$\mu(A) = \sum_k \mu(B^{(k)}),$$

which was to be proved.

In particular, the additivity of the elementary measures in Euclidean n-space follows from the additivity of the linear measure on the real line.

THEOREM 2. *If the measures $\mu_1, \mu_2, \cdots, \mu_n$ are σ-additive, then the measure $\mu_1 \times \cdots \times \mu_n$ is σ-additive.*

Proof. We carry out the proof for $n = 2$. Denote by λ_1 the Lebesgue extension of μ_1. Let $C = \bigcup_{n=1}^{\infty} C_n$, where C and C_n are in $\mathfrak{S}_1 \times \mathfrak{S}_2$, that is,

$$C = A \times B \qquad (A \in \mathfrak{S}_1, \quad B \in \mathfrak{S}_2),$$
$$C_n = A_n \times B_n \qquad (A_n \in \mathfrak{S}_1, \quad B_n \in \mathfrak{S}_2).$$

For $x \in A$ we set

$$f_n(x) = \begin{cases} \mu_2(B_n) & (x \in A_n), \\ 0 & (x \notin A_n). \end{cases}$$

It is easy to see that if $x \in A$,

$$\sum_n f_n(x) = \mu_2(B).$$

Consequently, in veiw of the Corollary to Theorem 2, §44,

$$\sum_n \int_A f_n(x) \, d\lambda_1 = \int_A \mu_2(B) \, d\mu_1(A) = \mu(C).$$

But

$$\int_A f_n(x) \, d\lambda_1 = \mu_2(B_n)\mu_1(A_n) = \mu(C_n),$$

so that

$$\sum_n \mu(C_n) = \mu(C).$$

The Lebesgue extension of the measure $\mu_1 \times \cdots \times \mu_n$ will be called the product of the measures μ_k and will be denoted by

$$\mu_1 \otimes \cdots \otimes \mu_n = \otimes_k \mu_k.$$

If

$$\mu_1 = \cdots = \mu_n = \mu,$$

we obtain the nth power of the measure μ:

$$\mu^n = \otimes_k \mu_k \qquad\qquad (\mu_k = \mu).$$

For instance, the n-dimensional Lebesgue measure μ^n is the nth power of the linear Lebesgue measure μ^1.

EXERCISES

1. If $\mathfrak{S}_1$ and $\mathfrak{S}_2$ are rings, then the collection of all finite disjoint unions of rectangles, i.e., elements of $\mathfrak{S}_1 \times \mathfrak{S}_2$, is a ring.

2. If $\mathfrak{S}_1$ and $\mathfrak{S}_2$ are rings each containing at least two distinct nonempty sets, then $\mathfrak{S}_1 \times \mathfrak{S}_2$ is not a ring.

3. Let $X = Y = [0, 1]$, let $\mathfrak{M}_1 = \mathfrak{M}_2$ be the collection of Lebesgue measurable sets, and let $\mu_1 = \mu_2$ be linear Lebesgue measure. The product measure $\mu = \mu_1 \times \mu_2$ on $\mathfrak{M}_1 \times \mathfrak{M}_2$ is not complete (see the end of §39). Hint: For $y \in Y$, $\mu(X \times y) = 0$. X contains a nonmeasurable subset M.

§47. The representation of plane measure in terms of the linear measure of sections, and the geometric definition of the Lebesgue integral

Let G be a region in the (x, y)-plane bounded by the verticals $x = a$, $x = b$ and by the curves $y = \varphi(x)$, $y = \psi(x)$.

The area of the region G is

$$V(G) = \int_a^b \{\varphi(x) - \psi(x)\} \, dx,$$

where the difference $\varphi(x_0) - \psi(x_0)$ is equal to the length of the section of the region G by the vertical $x = x_0$. Our problem is to carry over this method of measuring areas to an arbitrary product-measure

$$\mu = \mu_x \otimes \mu_y.$$

We shall assume that the measures μ_x and μ_y, defined on Borel algebras with units X and Y, respectively, are σ-additive and complete (if $B \subseteq A$ and $\mu(A) = 0$, then B is measurable). It was shown previously that all Lebesgue extensions have these properties.

We introduce the following notation:

$$A_x = \{y: (x, y) \in A\},$$
$$A_y = \{x: (x, y) \in A\}.$$

If X and Y are both real lines (so that $X \times Y$ is the plane), then A_{x_0} is the projection on the Y-axis of the section of the set A with the vertical $x = x_0$.

THEOREM 1. *Under the above assumptions,*

$$\mu(A) = \int_X \mu_y(A_x) \, d\mu_x = \int_Y \mu_x(A_y) \, d\mu_y$$

for an arbitrary μ-measurable set A.

(We note that integration over X actually reduces to integration over the set $U_y A_y \subset X$, in whose complement the function under the integral sign is zero. Similarly, $\int_Y = \int_B$, where $B = U_x A_x$.)

Proof. It is clearly sufficient to prove that

$$(1) \qquad \mu(A) = \int_X \varphi_A(x) \, d\mu_x,$$

where $\varphi_A(x) = \mu_y(A_x)$, since the second part of the theorem is completely analogous to the first. We note that the theorem includes the assertion that the set A_x is μ_y-measurable for almost all x (in the sense of the measure μ_x), and that the function $\varphi_A(x)$ is μ_x-measurable. If this were not so, (1) would have no meaning.

The measure μ, the Lebesgue extension of

$$m = \mu_x \times \mu_y,$$

is defined on the collection S_m of sets of the form

$$A = A_{y_0} \times A_{x_0},$$

where A_{y_0} is μ_x-measurable and A_{x_0} is μ_y-measurable.

Relation (1) is obvious for such sets, since

$$\varphi_A(x) = \begin{cases} \mu_y(A_{x_0}) & (x \in A_{y_0}), \\ 0 & (x \notin A_{y_0}). \end{cases}$$

Relation (1) can be extended without difficulty also to the sets of $\Re(S_m)$, that is, to finite unions of disjoint sets of S_m.

The proof of (1) in the general case is based on the following lemma, which has independent interest for the theory of Lebesgue extensions.

LEMMA. *If A is a μ-measurable set, there exists a set B such that*

$$B = \bigcap_n B_n, \qquad B_1 \supseteq B_2 \supseteq \cdots \supseteq B_n \supseteq \cdots,$$

$$B_n = \bigcup_k B_{nk}, \qquad B_{n1} \subseteq B_{n2} \subseteq \cdots \subseteq B_{nk} \subseteq \cdots,$$

where the sets B_{nk} are elements of $\mathfrak{R}(S_m)$, $A \subseteq B$ and

(2) $$\mu(A) = \mu(B).$$

Proof. The proof is based on the fact that, according to the definition of measurability, for arbitrary n the set A can be included in a union

$$C_n = \bigcup_r \Delta_{nr}$$

of sets Δ_{nr} of S_m such that $\mu(C_n) < \mu(A) + 1/n$.

Setting $B_n = \bigcap_{k=1}^n C_k$, it is easily seen that the sets B_n will have the form $B_n = \bigcup_s \delta_{ns}$, where the sets δ_{ns} are elements of S_m. Finally, putting

$$B_{nk} = \bigcup_{s=1}^k \delta_{ns},$$

we obtain the sets required by the lemma.

Relation (1) is easily extended with the aid of the sets $B_{nk} \in \mathfrak{R}(S_m)$ to the sets B_n and B by means of Theorem 2, §44, since

$$\varphi_{B_n}(x) = \lim_{k \to \infty} \varphi_{B_{nk}}(x), \qquad \varphi_{B_{n1}} \leq \varphi_{B_{n2}} \leq \cdots,$$

$$\varphi_B(x) = \lim_{n \to \infty} \varphi_{B_n}(x), \qquad \varphi_{B_1} \geq \varphi_{B_2} \geq \cdots.$$

If $\mu(A) = 0$, then $\mu(B) = 0$, and

$$\varphi_B(x) = \mu_y(B_x) = 0$$

a.e. Since $A_x \subseteq B_x$, A_x is measurable for almost all x and

$$\varphi_A(x) = \mu_y(A_x) = 0,$$

$$\int \varphi_A(x) \, d\mu_x = 0 = \mu(A).$$

Consequently, relation (1) holds for sets A such that $\mu(A) = 0$. If A is arbitrary, we write it as $A = B \setminus C$, where, in view of (2),

$$\mu(C) = 0.$$

Since (1) holds for B and C, it is easy to see that it also holds for A.

This completes the proof of Theorem 1.

We now consider the special case when Y is the real line, μ_y is linear Lebesgue measure and A is the set of points (x, y) such that

(3)
$$\begin{cases} x \in M, \\ 0 \le y \le f(x), \end{cases}$$

where M is a μ_x-measurable set and $f(x)$ is an integrable nonnegative function. Then

$$\mu_y(A_x) = \begin{cases} f(x) & (x \in M), \\ 0 & (x \in M), \end{cases}$$

$$\mu(A) = \int_M f(x)\, d\mu_x.$$

We have proved the following

THEOREM 2. *The Lebesgue integral of a nonnegative integrable function $f(x)$ is equal to the measure $\mu = \mu_x \otimes \mu_y$ of the set A defined by (3).*

If X is also the real line, the set M a closed interval and the function $f(x)$ Riemann integrable, this theorem reduces to the usual expression for the integral as the area under the graph of the function.

EXERCISES

1. The assumption that $\mu_x(X) < \infty$ and $\mu_y(Y) < \infty$, or more generally that X and Y are countable unions of sets of finite measure cannot be dropped from Theorem 1. Let $X = Y = [0, 1]$, let $\mathfrak{S}_x$ be the class of Lebesgue measurable sets of X, μ_x Lebesgue measure, $\mathfrak{S}_y$ the class of all subsets of Y, $\mu_y(A)$ the number of points in A, $A \subseteq Y$. If $E = \{(x, y):x = y\}$, show that $\int_X \mu_y(E_x)\, d\mu_x = 1$, but $\int_Y \mu_x(E_y)\, d\mu_y = 0$.

2. Under the hypotheses of Theorem 2, the graph of a nonnegative measurable function, i.e., $\{(x, f(x)):x \in M\}$, has μ-measure zero.

3. Let $X = Y = [0, 1]$, let $\mu_x = \mu_y$ be linear Lebesgue measure and set $\mu = \mu_x \otimes \mu_y$. Suppose that $A \subseteq X$ is nonmeasurable and that $B \subseteq Y$ is such that $\mu_y(B) = 0$.

a) $\chi_{A \times B}(x, y)$ is μ-measurable.

b) $\chi_{A \times B}(x, y)$ is μ_x-measurable for almost all (but not all) $y \in Y$.

4. If A and B are measurable subsets of $X \times Y$, i.e., $(\mu = \mu_x \otimes \mu_y)$-measurable, and $\mu_y(A_x) = \mu_y(B_x)$ for almost every $x \in X$, then

$$\mu(A) = \mu(B).$$

5. Suppose $v = f(u)$ is a strictly increasing continuous function defined on $[0, \infty)$ with $f(0) = 0$ and $\lim_{u \to \infty} f(u) = \infty$. Then $u = f^{-1}(v) = g(v)$ also has all these properties. Suppose that $0 \le u_0 < \infty$, $0 \le v_0 < \infty$, $U = [0, u_0]$, $V = [0, v_0]$, and $\mu_u = \mu_v$ is linear Lebesgue measure. Let

$$F(u_0) = \int_{[0, u_0]} f(u)\, d\mu_u, \quad G(v_0) = \int_{[0, v_0]} g(v)\, d\mu_v.$$

Prove Young's inequality:

$$u_0 v_0 \leq F(u_0) + G(v_0),$$

where the equality holds if, and only if, $v_0 = f(u_0)$, or equivalently

$$u_0 = g(v_0).$$

The result can be demonstrated as follows:

a) Let

$$E_1 = \{(u, v): 0 \leq v \leq f(u), \quad 0 \leq u \leq u_0\},$$
$$E_2 = \{(u, v): 0 \leq u \leq g(v), \quad 0 \leq v \leq v_0\},$$
$$I = U \times V, \quad \mu = \mu_u \otimes \mu_v.$$

Show that $I = (I \cap E_1) \cup (I \cap E_2)$, with $\mu[(I \cap E_1) \cap (I \cap E_2)] = 0$ (use §47, Ex. 2). It follows that

$$\mu I = u_0 v_0 = \mu(I \cap E_1) + \mu(I \cap E_2).$$

b) Show that

$$F(u_0) = \int_U f(u) \, d\mu_u = \int_U \left(\int_{[0, f(u)]} d\mu_v \right) d\mu_u$$
$$\geq \int_U \left(\int_{[0, \min(v_0, f(u))]} d\mu_v \right) d\mu_u$$
$$= \mu(I \cap E_1)$$

(use §47, Theorem 2).

Similarly, one shows that $G(u_0) \geq \mu(I \cap E_2)$.

The result is now clear.

§48. Fubini's theorem

Consider a triple product

(1) $$U = X \times Y \times Z.$$

We shall identify the point

$$(x, y, z) \in U$$

with the points

$$((x, y), z),$$
$$(x, (y, z))$$

of the products

(2) $$(X \times Y) \times Z,$$

(3) $$X \times (Y \times Z).$$

We therefore agree to regard the products (1), (2) and (3) as identical. If measures μ_x, μ_y, μ_z are defined on X, Y, Z, then the measure.

$$\mu_u = \mu_x \otimes \mu_y \otimes \mu_z$$

may be defined as

$$\mu_u = (\mu_x \otimes \mu_y) \otimes \mu_z,$$

or as

$$\mu_u = \mu_x \otimes (\mu_y \otimes \mu_z).$$

We omit a rigorous proof of the equivalence of these definitions, although it is not difficult.

We shall apply these general ideas to prove the fundamental theorem of the theory of multiple integrals.

FUBINI'S THEOREM. *Suppose that σ-additive and complete measures μ_x and μ_y are defined on Borel algebras with units X and Y, respectively; further, suppose that*

$$\mu = \mu_x \otimes \mu_y,$$

and that the function $f(x, y)$ is μ-integrable on

$$A = A_{y_0} \times A_{x_0}.$$

Then (see the parenthetical remark on p. 69)

(4)
$$\int_A f(x, y)\, d\mu = \int_X \left(\int_{A_x} f(x, y)\, d\mu_y \right) d\mu_x$$
$$= \int_Y \left(\int_{A_y} f(x, y)\, d\mu_x \right) d\mu_y.$$

Proof. The theorem includes among its assertions the existence of the integrals in parentheses for almost all values of the variables with respect to which the integrals are taken.

We shall prove the theorem first for the case $f(x, y) \geq 0$. To this end consider the triple product

$$U = X \times Y \times D^1,$$

where the third term is the real line, and the product measure

$$\lambda = \mu_x \otimes \mu_y \otimes \mu^1 = \mu \otimes \mu^1,$$

where μ^1 is linear Lebesgue measure.

We define a subset W of U as follows:

$$(x, y, z) \in W$$

if

$$x \in A_{y_0}, \qquad y \in A_{x_0},$$
$$0 \leq z \leq f(x, y).$$

In view of Theorem 2, §47,

$$(5) \qquad \lambda(W) = \int_A f(x, y) \, d\mu.$$

On the other hand, by Theorem 1, §47,

$$(6) \qquad \lambda(W) = \int_X \xi(W_x) \, d\mu_x,$$

where $\xi = \mu_y \otimes \mu^1$ and W_x is the set of pairs (y, z) for which $(x, y, z) \in W$. By Theorem 2, §47,

$$(7) \qquad \xi(W_x) = \int_{A_x} f(x, y) \, d\mu_y.$$

Comparing (5), (6), and (7), we obtain

$$\int_A f(x, x) \, d\mu = \int_X \left(\int_{A_x} f(x, y) \, d\mu_y \right) d\mu_x.$$

This completes the proof of the theorem if $f(x, y) \geq 0$. The general case is reduced to the case $f(x, y) \geq 0$ by means of the relations

$$f(x, y) = f^+(x, y) - f^-(x, y),$$

$$f^+(x, y) = \tfrac{1}{2}[|f(x, y)| + f(x, y)], \qquad f^-(x, y) = \tfrac{1}{2}[|f(x, y)| - f(x, y)].$$

REMARK. It can be shown that if $f(x, y)$ is μ-measurable, then

$$\int_A f(x, y) \, d\mu$$

exists if

$$\int_X \left(\int_{A_x} |f(x, y)| \, d\mu_y \right) d\mu_x$$

exists.

EXAMPLES where (4) does not hold.

1. Let

$$A = [-1, 1]^2,$$
$$(x, y) = xy/(x^2 + y^2)^2.$$

Then

$$\int_{-1}^{1} f(x, y) \, dx = 0 \qquad (y \neq 0),$$

and

$$\int_{-1}^{1} f(x, y) \, dy = 0 \qquad (x \neq 0).$$

Therefore

$$\int_{-1}^{1} \left(\int_{-1}^{1} f(x, y) \, dx \right) dy = \int_{-1}^{1} \left(\int_{-1}^{1} f(x, y) \, dy \right) dx = 0;$$

but the Lebesgue double integral over the square does not exist, since

$$\int_{-1}^{1} \int_{-1}^{1} |f(x, y)| \, dx \, dy \geq \int_{0}^{1} dr \int_{0}^{2\pi} (\sin \varphi \cos \varphi / r) \, d\varphi = 2 \int_{0}^{1} dr/r = \infty.$$

2. $A = [0, 1]^2$,

$$f(x, y) = \begin{cases} 2^{2n} & [(\tfrac{1}{2})^n \leq x < (\tfrac{1}{2})^{n-1}, \quad (\tfrac{1}{2})^n \leq y < (\tfrac{1}{2})^{n-1}], \\ -2^{2n+1} & [(\tfrac{1}{2})^{n+1} \leq x < (\tfrac{1}{2})^n, \quad (\tfrac{1}{2})^n \leq y < (\tfrac{1}{2})^{n-1}], \\ 0 & \text{(for all other points in the square).} \end{cases}$$

A simple calculation shows that

$$\int_{0}^{1} \left(\int_{0}^{1} f(x, y) \, dx \right) dy = 0, \qquad \int_{0}^{1} \left(\int_{0}^{1} f(x, y) \, dy \right) dx = 1.$$

EXERCISES

1. Suppose $f(x)$ and $g(y)$ are integrable over X and Y, respectively, and $h(x, y) = f(x)g(y)$. Show that $h(x, y)$ is $(\mu = \mu_x \otimes \mu_y)$-integrable over $X \times Y$ and

$$\int_{X \times Y} h(x, y) \, d\mu = \int_{X} f(x) \, d\mu_x \int_{Y} g(y) \, d\mu_y.$$

2. Suppose that $X = Y = [a, b]$ and that $\mu_x = \mu_y$ is linear Lebesgue measure. Let $f(x)$, $g(x)$ be integrable over X and periodic with period $b - a$. The *convolution* $f * g$ of f and g is also defined as a periodic function on $[a, b]$ by

$$(f * g)(x) = \int_{[a,b]} f(x - y)g(y) \, d\mu_y.$$

Show that $(f * g)(x)$ is integrable over X and

$$\int_X |(f*g)(x)| \, d\mu_x \leq \int_X |f(y)| \, d\mu_x \cdot \int_Y |g(y)| \, d\mu_y.$$

The integrability of $f * g$ demonstrates the existence of $(f * g)(x)$ for almost all $x \in X$. Hint: Use Fubini's theorem.

3. Suppose that $f(x)$ is integrable over $[0, b]$ with respect to Lebesgue measure. Suppose that $\alpha > 0$. The αth *fractional integral* of f is defined by

$$I_\alpha(f)(x) = [\Gamma(\alpha)]^{-1} \int_{[0,x]} (x - t)^{\alpha-1} f(t) \, d\mu_t$$

for $x \in [0, b]$, where $\Gamma(\alpha)$ is the Gamma function. Show that $I_\alpha(f)(x)$ is defined a.e. on $[0, b]$ and is integrable over $[0, a]$ for $a \in [0, b]$. Hint: Do directly as in Ex. 2, or use a suitable convolution of $f(x)$ with another function.

4. Suppose that $\alpha > 0$, $\beta > 0$, $f(x)$ is integrable on $[0, b]$ with respect to Lebesgue measure. By Ex. 3, $I_\beta[I_\alpha(f)(x)]$ is defined a.e. on $[0, b]$. Show that

$$I_\beta[I_\alpha(f)(x)] = I_{\alpha+\beta}(f)(x).$$

Hint: Use the result: for $p > 0$, $q > 0$,

$$\int_{[0,1]} x^{p-1}(1 - x)^{q-1} \, d\mu_x = \Gamma(p)\Gamma(q)/\Gamma(p + q).$$

5. (INTEGRATION BY PARTS.) Let $X = Y = [0, b]$ and let $\mu = \mu_x \otimes \mu_y$, where $\mu_x = \mu_y$ is linear Lebesgue measure.

Suppose that $f(x)$, $g(x)$ are integrable over X. If

$$F(x) = \int_{[0,x]} f(x) \, d\mu_x, \qquad G(x) = \int_{[0,x]} g(x) \, d\mu_x$$

for $x \in [0, 1]$, then

$$\int_X F(x)g(x) \, d\mu_x = F(b)G(b) - \int_X f(x)G(x) \, d\mu_x.$$

The result may be demonstrated as follows:

a) Let $E = \{(x, y):(x, y) \in X \times Y, y \leq x\}$. Show that E is μ-measurable. Hence χ_E is μ-measurable and $H(x, y) = \chi_E(x, y)g(x)f(y)$ is also μ-measurable.

b) Show that $H(x, y)$ is integrable over $X \times Y$ with respect to μ. (Apply Ex. 1.)

c) Apply Fubini's theorem to obtain

$$\int_X F(x)g(x)\, d\mu_x = \int_{X\times Y} H(x, y)\, d\mu = \int_Y f(y) \left(\int_{[y,b]} g(x)\, d\mu_x \right) d\mu_y .$$

This will yield the stated result.

§49. The integral as a set function

We shall consider the integral $F(A) = \int_A f(x)\, d\mu$ as a set function on the assumption that $S\mu$ is a Borel algebra with unit X and that $\int_X f(x)\, d\mu$ exists.

Then, as we have already proved:

1. $F(A)$ is defined on the Borel algebra S_μ.
2. $F(A)$ is real-valued.
3. $F(A)$ is additive, that is, if

$$A = \bigcup_n A_n \qquad\qquad (A, A_n \in S_\mu),$$

then

$$F(A) = \sum_n F(A_n).$$

4. $F(A)$ is absolutely continuous, that is, $\mu(A) = 0$ implies that

$$F(A) = 0.$$

We state the following important theorem without proof:

RADON'S THEOREM. *If a set function $F(A)$ has properties 1, 2, 3 and 4, it is representable in the form*

$$F(A) = \int_A f(x)\, d\mu.$$

We shall show that the function $f = dF/d\mu$ is uniquely defined a.e. In fact, if

$$F(A) = \int_A f_1(x)\, d\mu = \int_A f_2(x)\, d\mu$$

for all $A \in S_\mu$, then

$$\mu(A_n) \le n \int_{A_n} (f_1 - f_2)\, d\mu = 0$$

for arbitrary n, where

$$A_n = \{x : f_1(x) - f_2(x) > 1/n\}.$$

Similarly,

$$\mu(B_n) = 0$$

for

$$B_n = \{x : f_2(x) - f_1(x) > 1/n\}.$$

Since

$$\{x : f_1(x) \neq f_2(x)\} = \bigcup_n A_n \cup \bigcup_m B_m,$$

it follows that

$$\mu\{x : f_1(x) \neq f_2(x)\} = 0.$$

This proves our assertion.

EXERCISES

1. With the notation of this section, suppose that $f(x) \geq 0$ and let $\nu(A) = \int_A f(x) \, d\mu$. Then the conditions listed before Radon's theorem can be paraphrased by saying that $\nu(A)$ is a completely additive, absolutely continuous measure on the Borel algebra S_μ. Show that if $g(x)$ is integrable over X with respect to ν, then

$$\int_A g(x) \, d\nu = \int_A f(x) g(x) \, d\mu \qquad (A \in S_\mu).$$

2. If $\nu(A)$ is a completely additive measure on the Borel algebra S_μ, then ν may have the following property: For $\epsilon > 0$ there exists a $\delta > 0$ such that $A \in S_\mu$ and $\mu(A) < \delta$ imply $\nu(A) < \epsilon$. It is easy to see that if ν has this property, then ν is absolutely continuous with respect to μ, i.e., $\mu(A) = 0$ implies $\nu(A) = 0$. Show, conversely, that if ν is absolutely continuous with respect to μ, then ν has the above (ϵ, δ) property.

Chapter VIII

SQUARE INTEGRABLE FUNCTIONS

One of the most important linear normed spaces in Functional Analysis is Hilbert space, named after the German mathematician David Hilbert, who introduced this space in his research on the theory of integral equations. It is the natural infinite-dimensional analogue of Euclidean n-space. We became acquainted with one of the important realizations of Hilbert space in Chapter III—the space l_2, whose elements are the sequences

$$x = (x_1, \cdots, x_n, \cdots)$$

satisfying the condition

$$\sum_{n=1}^{\infty} x_n^2 < \infty.$$

We can now use the Lebesgue integral to introduce a second, in certain respects more convenient, realization of Hilbert space—the space of square integrable functions. In this chapter we consider the definition and fundamental properties of the space of square integrable functions and show that it is isometric (if certain assumptions are made about the measure used in the integral) to the space l_2.

We shall give an axiomatic definition of Hilbert space in Chapter IX.

§50. The space L_2

In the sequel we consider functions $f(x)$ defined on a set R, on which a measure $\mu(E)$ is prescribed, satisfying the condition $\mu(R) < \infty$. The functions $f(x)$ are assumed to be measurable and defined a.e. on R. We shall not distinguish between functions equivalent on R. For brevity, instead of $\int_R$ we write simply $\int$.

DEFINITION 1. We say that $f(x)$ is a *square integrable* (or *summable*) *function* on R if the integral

$$\int f^2(x) \, d\mu$$

exists (is finite). The collection of all square integrable functions is denoted by L_2.

The fundamental properties of such functions follow.

THEOREM 1. *The product of two square integrable functions is an integrable function.*

79

The proof follows immediately from the inequality

$$| f(x)g(x) | \leq \tfrac{1}{2}[f^2(x) + g^2(x)]$$

and the properties of the Lebesgue integral.

COROLLARY 1. *A square integrable function $f(x)$ is integrable.*

For, it is sufficient to set $g(x) \equiv 1$ in Theorem 1.

THEOREM 2. *The sum of two functions of L_2 is an element of L_2.*

Proof. Indeed,

$$[f(x) + g(x)]^2 \leq f^2(x) + 2 | f(x)g(x) | + g^2(x),$$

and Theorem 1 implies that the three functions on the right are summable.

THEOREM 3. *If $f(x) \in L_2$ and α is an arbitrary number, then $\alpha f(x) \in L_2$.*

Proof. If $f \in L_2$, then

$$\int [\alpha f(x)]^2 \, d\mu = \alpha^2 \int f^2(x) \, d\mu < \infty.$$

Theorems 2 and 3 show that a linear combination of functions of L_2 is an element of L_2; it is also obvious that the addition of functions and multiplication of functions by numbers satisfy the eight conditions of the definition of a linear space (Chapter III, §21). Hence L_2 is a linear space.

We now define an *inner* (or a *scalar*) product in L_2 by setting

$$(1) \qquad (f, g) = \int_R f(x)g(x) \, d\mu.$$

An inner product is a real-valued function of pairs of vectors of a linear space satisfying the following conditions:

1) $(f, g) = (g, f)$.
2) $(f_1 + f_2, g) = (f_1, g) + (f_2, g)$.
3) $(\lambda f, g) = \lambda(f, g)$.
4) $(f, f) > 0$ if $f \neq 0$.

The fundamental properties of the integral immediately imply that Conditions 1)–3) are satisfied by (1). Inasmuch as we have agreed not to distinguish between equivalent functions (so that, in particular, the null element in L_2 is the collection of all functions on R equivalent to $f \equiv 0$), Condition 4) is also satisfied (see the Corollary to Theorem 9, §43). We therefore arrive at the following

DEFINITION 2. The space L_2 is the Euclidean space (a linear space with an inner product) whose elements are the classes of equivalent square integrable functions; addition of the elements of L_2 and multiplication by scalars are defined in the way usual for functions and the inner product is defined by

(1)
$$(f, g) = \int f(x)g(x) \, d\mu.$$

The Schwarz inequality, which in this case has the form

(2)
$$\left(\int f(x)g(x) \, d\mu \right)^2 \leq \int f^2(x) \, d\mu \cdot \int g^2(x) \, d\mu,$$

is satisfied in L_2, as it is in every Euclidean space (see Ex. 3, §56). The same is true for the triangle inequality

(3)
$$\left\{ \int [f(x) + g(x)]^2 \, d\mu \right\}^{\frac{1}{2}} \leq \left\{ \int f^2(x) \, d\mu \right\}^{\frac{1}{2}} + \left\{ \int g^2(x) \, d\mu \right\}^{\frac{1}{2}}.$$

In particular, the Schwarz inequality yields the following useful inequality:

(4)
$$\left(\int f(x) \, d\mu \right)^2 \leq \mu(R) \int f^2(x) \, d\mu.$$

To introduce a norm into L_2 we set

(5)
$$\| f \| = (f, f)^{\frac{1}{2}} = \left[\int f^2(x) \, d\mu \right]^{\frac{1}{2}} \qquad (f \in L_2).$$

EXERCISE. Using the properties 1)–4) of the inner product, prove that the norm defined by (5) satisfies Conditions 1–3 of the definition of a norm in §21.

The following theorem plays an important part in many problems of analysis:

THEOREM 4. *The space L_2 is complete.*

Proof. a) Let $\{f_n(x)\}$ be a fundamental sequence in L_2, i.e.,

$$\| f_n - f_m \| \to 0 \qquad (n, m \to \infty).$$

Then there is a subsequence of indices $\{n_k\}$ such that

$$\| f_{n_k} - f_{n_{k+1}} \| \leq (\tfrac{1}{2})^k.$$

Hence, in view of inequality (4), it follows that

$$\int | f_{n_k}(x) - f_{n_{k+1}}(x) | \, d\mu \leq [\mu(R)]^{\frac{1}{2}} \left\{ \int [f_{n_k}(x) - f_{n_{k+1}}(x)]^2 \, d\mu \right\}^{\frac{1}{2}}$$

$$\leq (\tfrac{1}{2})^k [\mu(R)]^{\frac{1}{2}}.$$

This inequality and the Corollary to Theorem 2, §44 imply that the series

$$| f_{n_1}(x) | + | f_{n_2}(x) - f_{n_1}(x) | + \cdots$$

converges a.e. on R. Then the series

$$f_{n_1}(x) + [f_{n_2}(x) - f_{n_1}(x)] + \cdots$$

also converges a.e. on R to a function

(6) $$f(x) = \lim_{k \to \infty} f_{n_k}(x).$$

Hence, we have proved that if $\{f_n(x)\}$ is a fundamental sequence of functions in L_2, it contains an a.e. convergent subsequence.

b) We shall now show that the function $f(x)$ defined by (6) is an element of L_2 and that

(7) $$\| f_n(x) - f(x) \| \to 0 \qquad (n \to \infty).$$

For sufficiently large k and l,

$$\int [f_{n_k}(x) - f_{n_l}(x)]^2 \, d\mu < \epsilon.$$

In view of Theorem 3, §44, we may take the limit under the integral sign in this inequality as $l \to \infty$. We obtain

$$\int [f_{n_k}(x) - f(x)]^2 \, d\mu \leq \epsilon.$$

It follows that $f \in L_2$ and $f_{n_k} \to f$. But the convergence of a subsequence of a fundamental sequence to a limit implies that the sequence itself converges to the same limit. [Convergence here means the fulfillment of (7); in this connection see §51.] This proves the theorem.

EXERCISES

1. If we define the distance $d(f_1, f_2)$ in $L_2(R, \mu)$ as

$$d(f_1, f_2) = \| f_1 - f_2 \| = \left\{ \int [f_1(x) - f_2(x)]^2 \, dx \right\}^{\frac{1}{2}},$$

then d satisfies the axioms for a metric space (see vol. 1, §8). Furthermore, d is translation invariant, i.e.,

$$d(f_1 + f, f_2 + f) = d(f_1, f_2)$$

for $f_1, f_2, f \in L_2$. This result, of course, holds in any normed linear space (see vol. 1, §21).

2. Let $R = [0, 1]$ and let μ be linear Lebesgue measure. Show that $\{f : \| f \| \leq 1\}$ is closed and bounded, but not compact.

3. With the notation of Ex. 2, show that the set of continuous functions on $[0, 1]$ is a linear manifold in L_2, but is not a subspace, i.e., is not closed. (For the terminology, see §57.)

4. A measurable function $f(x)$ is said to be *essentially bounded* (on R) if there exists an $a > 0$ such that $|f(x)| \leq a$ a.e. on R. The number a is called an *essential upper bound* of f on R. For an essentially bounded function f, let $m = \inf\{a: a \text{ an essential upper bound of } f\}$. The number m is called the essential supremum of f: $m = \text{ess. sup } f$.

a) Show that ess. sup f is the smallest essential upper bound of f on R.

b) Let $L_\infty(R, \mu)$ be the collection of essentially bounded functions on R. If we put $\|f\| = \text{ess. sup } f$, show that L_∞ becomes a normed linear space.

5. Let $L_p(R, \mu)$, $p \geq 1$, be the set of measurable functions f defined on R for which $|f(x)|^p$ is integrable over R.

a) If a, b are real numbers, show that

$$|a + b|^p \leq 2^p(|a|^p + |b|^p).$$

(The condition $p \geq 1$ is essential here.)

b) Show then that $L_p(R, \mu)$ is a linear space, i.e., $f, g \in L_p$ implies that $f + g \in L_p$ and that $f \in L_p$ and a real imply $af \in L_p$. Define $\|f\|_p = \left[\int_R |f(x)|^p \, d\mu\right]^{1/p}$. We shall shortly see that L_p is a normed space with $\|f\|_p$ as norm.

6. a) Suppose $p > 1$. Define q by the equation $1/p + 1/q = 1$. p and q are called *conjugate exponents*. Let $v = f(u) = u^{p-1}$. Then $u = g(v) = v^{1/p-1}$. Verify that the hypotheses of Young's inequality (§47, Ex. 5) are satisfied and that $F(u) = u^p/p$, $G(v) = v^q/q$, and that therefore

$$uv \leq u^p/p + v^q/q,$$

with equality if, and only if, $u^p = v^q$.

b) (HÖLDER INEQUALITY.) Suppose $f \in L_p(R, \mu)$, $g \in L_q(R, \mu)$, with p, q conjugate exponents. Show that

$$f(x)g(x) \in L_1(R, \mu) = L$$

and

$$\left|\int_R f(x)g(x) \, d\mu\right| \leq \int_R |f(x)g(x)| \, d\mu$$

$$\leq \left(\int_R |f(x)|^p \, d\mu\right)^{1/p} \left(\int_R |g(x)|^q \, d\mu\right)^{1/q}$$

$$= \|f\|_p \|g\|_q.$$

This result may be obtained as follows: It is trivial if $\|f\|_p = 0$ or $\|g\|_q = 0$. Otherwise, put

$$u = |f(x)| / \|f\|_p, \qquad v = |g(x)| / \|g\|_q$$

in the result of a), and integrate over R (see vol. 1, p. 20).

c) (MINKOWSKI'S INEQUALITY.) If $f, g \in L_p(R, \mu)$, then

$$\|f + g\|_p \leq \|f\|_p + \|g\|_p,$$

or, in terms of integrals,

$$\left(\int_R |f(x) + g(x)|^p \, d\mu\right)^{1/p} \leq \left(\int_R |f(x)|^p \, d\mu\right)^{1/p} + \left(\int_R |g(x)|^p \, d\mu\right)^{1/p}.$$

If $\|f + g\|_p = 0$, then the result is clear. If $\|f + g\|_p > 0$, observe that

$$|f(x) + g(x)|^p \leq |f(x)| \, |f(x) + g(x)|^{p-1} + |g(x)| \, |f(x) + g(x)|^{p-1},$$
$$|f(x) + g(x)|^{p-1} \in L_q.$$

Apply Hölder's inequality to each term on the right to obtain

$$\int_R |f + g|^p \, d\mu \leq \left(\int_R |f + g|^p \, d\mu\right)^{1/q} (\|f\|_p + \|g\|_p).$$

It is now clear that $L_p(R, \mu)$ with norm $\|f\|_p$ is a normed linear space for $p > 1$. Note also that if $p = 2$, then $q = 2$, and Hölder's inequality reduces to the Schwarz inequality.

§51. Mean convergence. Dense subsets of L_2

The introduction of a norm in L_2 determines a new notion of convergence for square integrable functions:

$$f_n \to f \qquad\qquad (\text{in } L_2)$$

if

$$\lim_{n \to \infty} \int [f_n(x) - f(x)]^2 \, d\mu = 0.$$

This type of convergence of functions is called *mean convergence*, or, more precisely, mean square convergence.

Let us consider the relation of mean convergence to uniform convergence and convergence a.e. (see Chapter VI).

THEOREM 1. *If a sequence* $\{f_n(x)\}$ *of functions of* L_2 *converges uniformly to* $f(x)$, *then* $f(x) \in L_2$, *and* $\{f_n(x)\}$ *is mean convergent to* $f(x)$.

Proof. Suppose that $\epsilon > 0$. If n is sufficiently large,

$$|f_n(x) - f(x)| < \epsilon,$$

whence

$$\int [f_n(x) - f(x)]^2 \, d\mu < \epsilon^2 \mu(R).$$

The theorem follows at once from this inequality.

Theorem 1 implies that if an arbitrary $f \in L_2$ can be approximated with arbitrary accuracy by functions $f_n \in M \subseteq L_2$ in the sense of uniform convergence, then it can be approximated by such functions in the sense of mean convergence.

Hence an arbitrary function $f \in L_2$ can be approximated with arbitrary accuracy by *simple* functions belonging to L_2.

We prove that an arbitrary simple function $f \in L_2$, and consequently an arbitrary function of L_2, can be approximated to any desired degree of accuracy by simple functions whose set of distinct values is finite.

Suppose that $f(x)$ assumes the values $y_1, \cdots, y_n, \cdots$ on the sets $E_1, \cdots, E_n, \cdots$. Inasmuch as f^2 is summable, the series

$$\sum_n y_n^2 \mu(E_n) = \int f^2(x) \, d\mu$$

converges. Choose an N such that

$$\sum_{n>N} y_n^2 \mu(E_n) < \epsilon,$$

and set

$$f_N(x) = \begin{cases} f(x) & (x \in E_i, i \leq N), \\ 0 & (x \in E_i, i > N). \end{cases}$$

Then

$$\int [f(x) - f_N(x)]^2 \, d\mu = \sum_{n>N} y_n^2 \mu(E_n) < \epsilon,$$

that is, the function $f_N(x)$, which assumes a finite set of values, approximates the function f with arbitrary accuracy.

Let R be a metric space with a measure possessing the following property (which is satisfied in all cases of practical interest): all the open and closed sets of R are measurable, and

$$(*) \qquad \mu^*(M) = \inf \{\mu(G); M \subseteq G\}$$

for all $M \subseteq R$, where the lower bound is taken over all open sets G containing M. Then we have

THEOREM 2. *The set of all continuous functions on R is dense in L_2.*

Proof. In view of the preceding discussion, it is sufficient to prove that every simple function assuming a finite number of values is the limit, in the sense of mean convergence, of continuous functions. Furthermore, since every simple function assuming a finite set of values is a linear combination of characteristic functions $\chi_M(x)$ of measurable sets, it is enough to carry out the proof for such functions. Let M be a measurable set in the metric space R. Then it follows at once from the Condition (∗) that for every $\epsilon > 0$ there exists a closed set F_M and an open set G_M such that

$$F_M \subseteq M \subseteq G_M, \qquad \mu(G_M) - \mu(F_M) < \epsilon.$$

We now define

$$\varphi_\epsilon(x) = \rho(x, R \setminus G_M)/[\rho(x, R \setminus G_M) + \rho(x, F_M)].$$

This function is 0 on $R \setminus G_M$ and 1 on F_M. It is continuous, since $\rho(x, F_M)$, $\rho(x, R \setminus G_M)$ are continuous and their sum does not vanish. The function $\chi_M(x) - \varphi_\epsilon(x)$ is bounded by 1 on $G_M \setminus F_M$ and vanishes in the complement of this set. Consequently,

$$\int [\chi_M(x) - \varphi_\epsilon(x)]^2 \, d\mu < \epsilon,$$

and the theorem follows.

THEOREM 3. *If a sequence $\{f_n(x)\}$ converges to $f(x)$ in the mean, it contains a subsequence $\{f_{n_k}(x)\}$ which converges to $f(x)$ a.e.*

Proof. If $\{f_n(x)\}$ converges in the mean, it is a fundamental sequence in L_2. Therefore, repeating the reasoning in Part a) of the proof of Theorem 4, §50, we obtain a subsequence $\{f_{n_k}(x)\}$ of $\{f_n(x)\}$ which converges a.e. to a function $\varphi(x)$. Furthermore, Part b) of the same proof shows that $\{f_{n_k}(x)\}$ converges to $\varphi(x)$ in the mean. Hence, $\varphi(x) = f(x)$ a.e.

★ It is not hard to find examples to show that convergence in the mean does not imply convergence a.e. In fact, the sequence of functions $\{f_{n_k}\}$ defined on p. 45 obviously converges in the mean to $f \equiv 0$, but (as was shown) does not converge a.e. We shall now show that convergence a.e. (and even everywhere) does not imply mean convergence. Let

$$f_n(x) = \begin{cases} n & [x \in (0, 1/n)], \\ 0 & \text{for all remaining values of } x. \end{cases}$$

It is clear that the sequence $\{f_n(x)\}$ converges to 0 everywhere on $[0, 1]$, but that

$$\int_0^1 f_n^2(x) \, dx = n \to \infty.$$

The Chebyshev inequality (§43, Theorem 9) implies that if a sequence is mean convergent, it converges in measure. Therefore, Theorem 3, which we have proved in this section independently of the Chebyshev inequality, follows from Theorem 4, §41. The relations between the various types of convergence of functions can be schematized as follows:

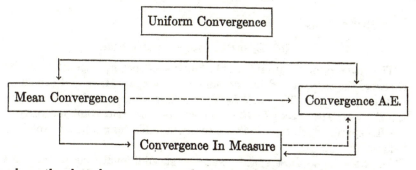

where the dotted arrows mean that a sequence converging in measure contains a subsequence converging a.e. and that a sequence converging in the mean contains a subsequence converging a.e. ★

EXERCISES

1. If $\{f_n(x)\}$ converges to $f(x)$ in the mean and $\{f_n(x)\}$ converges pointwise a.e. to $g(x)$, then $f(x) = g(x)$ a.e. on R.

2. a) If $\{f_n(x)\} \subset L_2$, $\{f_n(x)\}$ converges to $f(x)$ pointwise a.e. and $|f_n(x)| \leq g(x)$, $g \in L_2$, then $\{f_n(x)\}$ converges in the mean to $f(x)$, i.e., $\|f_n - f\|_2 \to 0$.

   b) The corresponding result obtains for L_p ($p > 1$).

3. a) If $\|f_n - f\|_2 \to 0$, then $\|f_n\|_2 \to \|f\|_2$; hence

$$\lim_{n \to \infty} \int_R |f_n(x)|^2 \, d\mu = \int_R |f(x)|^2 \, d\mu.$$

   b) The corresponding result obtains for L_p ($p > 1$).

4. a) Let $\{f_n(x)\}$ converge to $f(x)$ in the mean and suppose $g(x) \in L_2$. Then $\{f_n(x)g(x)\}$ converges to $f(x)g(x)$.

   b) More generally, if $\|f_n - f\|_2 \to 0$ and $\|g_n - g\|_2 \to 0$, then $\|f_n g_n - fg\|_2 \to 0$.

   c) Similar results obtain for $f_n, f \in L^p$; $g_n, g \in L^q$, with $1/p + 1/q = 1$, $p > 1$.

   d) Let $R = [a, b]$, and let μ be linear Lebesgue measure. Then $\{f_n\}$ converges to f in the mean implies that

$$\lim_{n \to \infty} \int_{[a,x_0]} f_n(x) \, d\mu = \int_{[a,x_0]} f(x) \, d\mu \qquad (a \leq x_0 \leq b).$$

Hint: Choose

$$g(x) = \begin{cases} 1 & (a \le x < x_0), \\ 0 & (x > x_0). \end{cases}$$

Show that $g \in L_2$.

§52. L_2 spaces with countable bases

The space L_2 of square integrable functions depends, in general, on the choice of the space R and the measure μ. To designate it fully it should be written as $L_2(R, \mu)$. The space $L_2(R, \mu)$ is finite-dimensional only in exceptional cases. The spaces $L_2(R, \mu)$ which are most important for analysis are the spaces which have *infinite dimension* (this term will be explained below).

To characterize these spaces, we need an additional concept from the theory of measure.

We can introduce a metric in the collection $\mathfrak{M}$ of measurable subsets of the space R (whose measure we have assumed to be finite) by setting

$$\rho(A, B) = \mu(A \triangle B).$$

If we identify sets A and B for which $\mu(A \triangle B) = 0$ (that is, we consider sets which are the same except for a set of measure zero to be indistinguishable), then the set $\mathfrak{M}$ together with the metric ρ becomes a metric space.

DEFINITION. A measure μ is said to have a *countable base* if the metric space $\mathfrak{M}$ contains a countable dense set.

In other words, a measure μ has a countable base if there is a countable set

$$\mathfrak{D} = \{A_n\} \qquad (n = 1, 2, \cdots)$$

of measurable subsets of R (a countable base for the measure μ) such that for every measurable $M \subseteq R$ and $\epsilon > 0$ there is an $A_k \in \mathfrak{D}$ for which

$$\mu(M \triangle A_k) < \epsilon.$$

In particular, a measure μ obviously has a countable base if it is the extension of a measure defined on a countable collection S_m of sets. Indeed, in that case the ring $\Re(S_m)$ (which is obviously countable) is the required base, in view of Theorem 3, §38.

In particular, Lebesgue measure on a closed interval of the real line is induced by the set of intervals with rational endpoints as elementary sets. Since the collection of all such intervals is countable, Lebesgue measure has a countable base.

The product $\mu = \mu_1 \otimes \mu_2$ of two measures with countable bases also has a countable base, since, as is easily seen, the finite unions of the products of pairs of elements of the bases of μ_1 and μ_2 form a base for $\mu = \mu_1 \otimes \mu_2$. Therefore, Lebesgue measure in the plane (as well as in n-dimensional space) has a countable base.

Suppose that

$$(1) \qquad A_1^*, \cdots, A_n^*, \cdots$$

is a countable base for μ. It is easy to see that the base (1) can be extended to a base

$$(2) \qquad A_1, \cdots, A_n, \cdots$$

of μ with the following properties:

1) The base (2) is closed under differences.
2) R is an element of the base (2).

Conditions 1) and 2) imply that (2) is closed under finite unions and intersections. This follows from the following relations:

$$A_1 \cap A_2 = A_1 \setminus (A_1 \setminus A_2),$$
$$A_1 \cup A_2 = R \setminus [(R \setminus A_1) \cap (R \setminus A_2)].$$

THEOREM. *If a measure μ has a countable base, then $L_2(R, \mu)$ contains a countable dense set.*

Proof. The finite sums

$$(3) \qquad \sum_{k=1}^n c_k f_k(x),$$

where the c_k are rational numbers and the $f_k(x)$ are characteristic functions of the elements of the countable base of μ, form the required base for $L_2(R, \mu)$.

For, as we have already shown in the preceding section, the set of simple functions assuming only a finite number of values, is everywhere dense in L_2. Since it is obvious that an arbitrary function of this set can be approximated with arbitrary accuracy by functions of the same form, but assuming only rational values, and since the set of functions of the form (3) is countable, to prove the theorem it is sufficient to show that an arbitrary simple function $f(x)$, assuming the values

$$y_1, \cdots, y_n \qquad\qquad (y_i \text{ rational})$$

on the sets

$$E_1, \cdots, E_n \qquad [\cup_i E_i = R, E_i \cap E_j = \emptyset \ (i \neq j)],$$

can be approximated with arbitrary accuracy by functions of the form (3) in the sense of the metric in L_2. In view of this remark, we may assume without loss of generality that the base for μ satisfies 1) and 2).

By definition, for every $\epsilon > 0$ there exist sets $A_1, \cdots, A_n$ of the base for μ such that $\rho(E_k, A_k) < \epsilon$, that is,

$$\mu[(E_k \setminus A_k) \cup (A_k \setminus E_k)] < \epsilon.$$

We set

$$A_1' = A_1,$$
$$A_k' = A_k \setminus \bigcup_{i<k} A_i \qquad\qquad (2 \le k \le n),$$

and define

$$f^*(x) = \begin{cases} y_k & (x \in A_k'), \\ 0 & (x \in R \setminus \bigcup_{i=1}^{n} A_i'). \end{cases}$$

It is easy to see that

$$\mu\{x : f(x) \ne f^*(x)\}$$

is arbitrarily small for sufficiently small ϵ, and that consequently the integral

$$\int [f(x) - f^*(x)]^2 \, d\mu \le (2 \max |y_n|)^2 \, \mu\{x : f(x) \ne f^*(x)\}$$

is arbitrarily small for sufficiently small ϵ.

In view of our assumptions about the base of μ, $f^*(x)$ is of the form (3). This proves the theorem.

If R is a closed interval on the real line and μ is Lebesgue measure, a countable base in $L_2(R, \mu)$ can be obtained in a more classical way: for instance, the set of all polynomials with rational coefficients forms a base in L_2. This set is dense (even in the sense of uniform convergence) in the set of continuous functions, and the latter are a dense set in $L_2(R, \mu)$.

In the sequel we restrict ourselves to spaces $L_2(R, \mu)$ with countable dense subsets [that is, separable spaces (see §9)].

EXERCISES

1. Let X be the unit square in the plane, μ the σ-additive measure that is the Lebesgue extension of the measure defined in §35, Ex. 2 (see also §37, Ex. 3). Note that to obtain the Lebesgue extension we must use the method of §39 since the semi-ring of horizontal cells has no unit.

Show that μ is not separable.

Hint: Any $A \in S_\mu$ is a countable union of horizontal line segments; hence, any countable collection of elements of S_μ contains a countable set of horizontal line segments $\{A_n\}$. We can choose $y_0 \in [0, 1]$ such that

$$A_0 = \{(x, y): 0 \le x \le 1, y = y_0\}$$

is disjoint from all these. Then show that

$$\rho(A_0, A_n) = \mu(A_0 \, \Delta \, A_n) \ge 1$$

for all n.

2. a) Since $\mu(R) < \infty$, a simple measurable function with only finitely many distinct values belongs to $L_p(R, \mu)$ for every $p \ge 1$.

b) The discussion in §51 can be paralleled for the present case to show that every $f \in L_p$ can be approximated in the mean of order p (i.e., in the metric of L_p) by simple functions assuming only finitely many values.

c) The procedure of §52 can now be imitated to show that if μ has a countable base, then $L_p(R, \mu)$ is separable for $p \ge 1$.

d) It follows from a) and c) that for $r, s \ge 1$, $L_r(R, \mu)$ is dense in $L_s(R, \mu)$.

§53. Orthogonal sets of functions. Orthogonalization

In this section we consider functions $f \in L_2$ defined on a measurable set R with measure μ, which we assume to have a countable base and to satisfy the condition $\mu(R) < \infty$. As before, we do not distinguish between equivalent functions.

DEFINITION 1. A set of functions

$$(1) \qquad \varphi_1(x), \, \cdots, \, \varphi_n(x)$$

is said to be *linearly dependent* if there exist constants $c_1, \, \cdots, \, c_n$, not all zero, such that

$$(2) \qquad c_1\varphi_1(x) + c_2\varphi_2(x) + \cdots + c_n\varphi_n(x) = 0$$

a.e. on R. If, however, (2) implies that

$$(3) \qquad c_1 = \cdots = c_n = 0,$$

then the set (1) is said to be *linearly independent*.

Clearly, a linearly independent set cannot contain functions equivalent to $\psi(x) \equiv 0$.

DEFINITION 2. An infinite sequence of functions

$$(4) \qquad \varphi_1(x), \, \cdots, \, \varphi_n(x), \, \cdots$$

is said to be linearly independent if every finite subset of (4) is linearly independent.

We denote by

$$M = M(\varphi_1, \, \cdots, \, \varphi_n, \, \cdots) = M\{\varphi_k\}$$

the set of all finite linear combinations of functions of (4). This set is called the *linear hull* of (4) (or the *linear manifold generated by* $\{\varphi_k\}$). We let

$$\bar{M} = \bar{M}(\varphi_1, \cdots, \varphi_n, \cdots) = \bar{M}\{\varphi_k\}$$

denote the closure of M in L_2. $\bar{M}$ is called the *closed linear hull* of (4) (or the *subspace generated by* $\{\varphi_k\}$).

It is easily seen that $\bar{M}$ consists precisely of the functions $f \in L_2$ which can be approximated by finite linear combinations of functions of (4) with arbitrarily prescribed accuracy.

DEFINITION 3. The set of functions (4) is said to be *complete* (sometimes *closed*) if

$$\bar{M} = L_2.$$

Suppose that L_2 contains a countable dense set of functions

$$f_1, \cdots, f_n, \cdots.$$

Deleting from this set those functions which are linearly dependent on the preceding functions in the sequence, we obtain a linearly independent set of functions

$$g_1, \cdots, g_n, \cdots$$

which, as is easily seen, is complete.

If L_2 contains a finite complete set (1) of linearly independent functions, then

$$L_2 = \bar{M}(\varphi_1, \cdots, \varphi_n) = M(\varphi_1, \cdots, \varphi_n)$$

is isomorphic to Euclidean n-space. We say L_2 is n-dimensional. Otherwise, we call L_2 infinite-dimensional.

The space L_2 is infinite-dimensional in all cases of interest in analysis.

Obviously, in order that (4) be complete it is sufficient that it be possible to approximate every function of a dense subset of L_2 with arbitrary accuracy by linear combinations of functions of (4).

Let $R = [a, b]$ be a closed interval on the real line with the usual Lebesgue measure. Then the set of functions

(5) $$1, x, x^2, \cdots, x^n, \cdots$$

is complete in $L_2(R, \mu)$.

For, according to the Weierstrass theorem (see Vol. 1, p. 25) the set of linear combinations of functions (5) is dense in the set of all continuous functions. The completeness of the set (5) now follows from this remark and Theorem 2, §51.

Two functions $f(x)$ and $g(x)$ of L_2 are said to be *orthogonal* if

$$(f, g) = \int f(x)g(x)\, d\mu = 0.$$

We shall call every set of functions $\varphi_1, \cdots, \varphi_n, \cdots$, which are distinct from zero and are pairwise orthogonal, an *orthogonal set*. An orthogonal set is said to be *normalized* or *orthonormal* if $\| \varphi_n \| = 1$ for all n; in other words,

$$\varphi_1, \cdots, \varphi_n, \cdots$$

is an orthonormal set of functions if

$$(\varphi_i, \varphi_k) = \int \varphi_i(x)\varphi_k(x)\, d\mu = \begin{cases} 0 & (i \neq k), \\ 1 & (i = k). \end{cases}$$

EXAMPLES: 1. A classical example of an orthonormal set of functions on the closed interval $[-\pi, \pi]$ is the set of trigonometric functions

$$(2\pi)^{-\frac{1}{2}},\ (\pi)^{-\frac{1}{2}} \cos x,\ (\pi)^{-\frac{1}{2}} \cos 2x,\ \cdots,\ (\pi)^{-\frac{1}{2}} \sin x,\ (\pi)^{-\frac{1}{2}} \sin 2x,\ \cdots.$$

2. The polynomials

$$P_n(x) = (2^n n!)^{-1}\, (d^n[(x^2 - 1)^n]/dx^n) \qquad (n = 0, 1, 2, \cdots),$$

called the *Legendre polynomials*, form an orthogonal set of functions on $[-1, 1]$. An orthonormal set consists of the functions

$$[\tfrac{1}{2}(2n + 1)]^{\frac{1}{2}} P_n(x).$$

It is easily seen that *an orthonormal set of functions is linearly independent.* For, multiplying the relation

$$c_1\varphi_1 + \cdots + c_n\varphi_n = 0$$

by φ_i and integrating, we obtain

$$c_i(\varphi_i, \varphi_i) = 0,$$

and since $(\varphi_i, \varphi_i) > 0$, $c_i = 0$.

We note further, that *if the space L_2 contains a countable dense set* $f_1, \cdots, f_n, \cdots$, *then an arbitrary orthonormal set of functions* $\{\varphi_\alpha\}$ *is at most countable.*

To see this, suppose that $\alpha \neq \beta$. Then

$$\| \varphi_\alpha - \varphi_\beta \| = 2^{\frac{1}{2}}.$$

For every α choose an f_α from the dense subset such that

$$\| \varphi_\alpha - f_\alpha \| < 2^{-\frac{1}{2}}.$$

Clearly, $f_\alpha \neq f_\beta$ if $\alpha \neq \beta$. Since the set of all f_α is countable, the set of φ_α is at most countable.

An orthonormal base is of great importance in studying finite-dimensional spaces. In this connection an orthonormal base is a set of orthogonal unit vectors, whose linear hull coincides with the whole space. In the infinite-dimensional case the analogue of such a base is a complete orthonormal set of functions, that is, a set

$$\varphi_1, \cdots, \varphi_n, \cdots$$

such that
   1) $(\varphi_i, \varphi_k) = \delta_{ik}$,
   2) $\bar{M}(\varphi_1, \cdots, \varphi_n, \cdots) = L_2$.
We gave examples of orthonormal sets of functions on the intervals $[-\pi, \pi]$ and $[-1, 1]$ above. The existence of a complete orthonormal set of functions in an arbitrary separable space L_2 is a consequence of the following theorem:

THEOREM. *Suppose that the set of functions*

$$(6) \qquad\qquad f_1, \cdots, f_n, \cdots$$

is linearly independent. Then there exists a set of functions

$$(7) \qquad\qquad \varphi_1, \cdots, \varphi_n, \cdots$$

satisfying the following conditions:
   1) *The set (7) is orthonormal.*
   2) *Every function φ_n is a linear combination of the functions $f_1, \cdots, f_n$:*

$$\varphi_n = a_{n1}f_1 + a_{n2}f_2 + \cdots + a_{nn}f_n,$$

with $a_{nn} \neq 0$.
   3) *Every function f_n is a linear combination of the functions*

$$f_n = b_{n1}\varphi_1 + \cdots + b_{nn}\varphi_n,$$

with $b_{nn} \neq 0$.

Every function of the set (7) is uniquely determined (except for sign) by the Conditions 1)–3).

Proof. The function $\varphi_1(x)$ is uniquely determined (except for sign) by the conditions of the theorem. For,

$$\varphi_1 = a_{11}f_1,$$

$$(\varphi_1, \varphi_1) = a_{11}{}^2(f_1, f_1) = 1,$$

whence

$$b_{11} = 1/a_{11} = (f_1, f_1)^{\frac{1}{2}}, \qquad \varphi_1 = \pm(f_1, f_1)^{-\frac{1}{2}}f_1.$$

Suppose that the functions φ_k ($k < n$) satisfying 1)–3) have already been determined. Then f_n may be written as

$$f_n = b_{n1}\varphi_1 + \cdots + b_{n,n-1}\,\varphi_{n-1} + h_n,$$

where $(h_n, \varphi_k) = 0$ ($k < n$).

Obviously, $(h_n, h_n) > 0$ [the assumption that $(h_n, h_n) = 0$ would contradict the linear independence of the set (6)].

Set

$$\varphi_n = (h_n, h_n)^{-\frac{1}{2}} h_n.$$

Then

$$(\varphi_n, \varphi_i) = 0 \qquad\qquad (i < n),$$

$$(\varphi_n, \varphi_n) = 1,$$

$$f_n = b_{n1}\varphi_1 + \cdots + b_{nn}\varphi_n \quad [b_{nn} = (h_n, h_n)^{\frac{1}{2}} \neq 0],$$

that is, the functions $\varphi_n(x)$ satisfy the conditions of the theorem. The last assertion of the theorem is an immediate consequence of the linear independence of the set $f_1, \cdots, f_n$.

The transition from a set (6) to the set (7) satisfying 1)–3) is called an *orthogonalization process*.

Obviously,

$$M(f_1, \cdots, f_n, \cdots) = M(\varphi_1, \cdots, \varphi_n, \cdots);$$

hence, the sets (6) and (7) are either both complete or not complete.

Therefore, the set (6) may be replaced by the set (7) in all problems of approximating functions f by linear combinations of the functions (6).

We said above that the existence in L_2 of a countable dense set implies the existence of a countable complete set of linearly independent functions. Orthogonalization of the latter set yields a complete countable orthonormal set.

EXERCISES

1. With the notation of the theorem in this section,

$$f_n = b_{n1}\varphi_1 + \cdots + b_{nn}\varphi_n.$$

Show that

$$b_{ni} = (f_n, \varphi_i) \qquad\qquad (1 \le i \le n).$$

2. Show that the set of functions

$$1, x, \cdots, x^n, \cdots$$

is linearly independent on any interval $[a, b]$.

3. For $R = [0, 1]$ and μ linear Lebesgue measure, show that $L_2(R, \mu)$ is infinite-dimensional.

4. Suppose that $\{f_1(x), \cdots, f_n(x)\} \subset L_2$. The Grammian of $\{f_i\}$ is the determinant

$$G_n = \det (f_i, f_j) \qquad (1 \leq i, j \leq n).$$

Show that $\{f_i\}$ is linearly dependent if, and only if, $G_n = 0$. Hints: Suppose $\{f_i\}$ is linearly dependent. Multiply the dependency relation by f_i $(1 \leq i \leq n)$ and integrate to obtain a system of homogeneous equations with a nontrivial solution. Conversely, if $G_n = 0$, then the same system has a nontrivial solution $a_1, \cdots, a_n$. Show that $(\sum_i a_i f_i, \sum_i a_i f_i) = 0$.

5. a) Show that the Legendre functions $P_n(x)$ given in the text form an orthogonal set.

b) Show that $\| P_n(x) \| = 2^{\frac{1}{2}}(2n + 1)^{-\frac{1}{2}}$. Hint: P_n is a polynomial of degree n. Use integration by parts repeatedly to show that $P_n(x)$ is orthogonal to x^k $(0 \leq k < n)$.

§54. Fourier series over orthogonal sets. The Riesz-Fisher theorem

If $e_1, e_2, \cdots, e_n$ is an orthonormal base in Euclidean n-space $R^{(n)}$, then every vector $x \in R^{(n)}$ can be written in the form

$$(1) \qquad x = \sum_{k=1}^n c_k e_k,$$

with

$$c_k = (x, e_k).$$

The purpose of this section is, in a sense, to generalize (1) to the infinite-dimensional case.

Let

$$(2) \qquad \varphi_1, \cdots, \varphi_n, \cdots$$

be an orthonormal set and suppose that $f \in L_2$.

We pose the following problem: For prescribed n determine the coefficients α_k $(1 \leq k \leq n)$ so that the distance, in the sense of the metric in L_2, between f and the sum

$$(3) \qquad S_n = \sum_{k=1}^n \alpha_k \varphi_k$$

is the least possible.

Set $c_k = (f, \varphi_k)$. Since the set (2) is orthonormal,

$$(4) \quad \| f - S_n \|^2 = (f - \sum_{k=1}^n \alpha_k \varphi_k, f - \sum_{k=1}^n \alpha_k \varphi_k)$$
$$= (f, f) - 2(f, \sum_{k=1}^n \alpha_k \varphi_k) + (\sum_{k=1}^n \alpha_k \varphi_k, \sum_{j=1}^n \alpha_j \varphi_j)$$

$$= \|f\|^2 - 2\sum_{k=1}^{n} \alpha_k c_k + \sum_{k=1}^{n} \alpha_k^2$$
$$= \|f\|^2 - \sum_{k=1}^{n} c_k^2 + \sum_{k=1}^{n} (\alpha_k - c_k)^2.$$

It is clear that the minimum of (4) is assumed when the last term is zero, i.e., if

(5) $$\alpha_k = c_k \qquad (1 \le k \le n).$$

In that case

(6) $$\|f - S_n\|^2 = (f, f) - \sum_{k=1}^{n} c_k^2.$$

DEFINITION. The numbers

$$c_k = (f, \varphi_k)$$

are called the *Fourier coefficients* of the function $f \in L_2$ relative to the orthonormal set (2), and the series

$$\sum_{k=1}^{\infty} c_k \varphi_k$$

(which may or may not converge) is called the *Fourier series* of the function f with respect to the set (2).

We have proved that of all the sums of the form (3) the partial sums of the Fourier series of the function deviate least (in the sense of the metric in L_2), for prescribed n, from the function f. The geometric meaning of this result may be explained as follows: The functions

$$f - \sum_{k=1}^{n} \alpha_k \varphi_k$$

are orthogonal to all the linear combinations of the form

$$\sum_{k=1}^{n} \beta_k \varphi_k ,$$

that is, these functions are orthogonal to the subspace generated by the functions $\varphi_1, \cdots, \varphi_n$ if, and only if, (5) is satisfied. (Verify this!) Hence, our result is a generalization of the well known theorem of elementary geometry which states that the length of the perpendicular from a given point to a line or a plane is less than that of any other line from the point to the given line or plane.

Since $\|f - S_n\|^2 \ge 0$, relation (4) implies that

$$\sum_{k=1}^{n} c_k^2 \le \|f\|^2,$$

where n is arbitrary and the right side is independent of n. Hence, the series $\sum_{k=1}^{\infty} c_k^2$ converges, and

(7) $$\sum_{k=1}^{\infty} c_k^2 \le \|f\|^2.$$

This is the *Bessel inequality*.

We introduce the following important

DEFINITION. An orthonormal set is said to be *closed* (sometimes *complete*) if

$$(8) \qquad \sum_{k=1}^{\infty} c_k^{\,2} = \| f \|^2$$

for every $f \in L_2$. The relation (8) is called *Parseval's equality*.

It is clear from (6) that the set (2) is closed if, and only if, the partial sums of the Fourier series of every function $f \in L_2$ converge to f in the metric of L_2 (that is, in the mean).

The notion of a closed orthonormal set is intimately related to the completeness of a set of functions (see §53).

THEOREM 1. *In L_2 every complete orthonormal set is closed, and conversely.*

Proof. Suppose that $\{\varphi_n(x)\}$ is closed; then the sequence of partial sums of the Fourier series of every $f \in L_2$ is mean convergent. Hence the linear combinations of the elements of the set $\{\varphi_n(x)\}$ are dense in L_2, that is, $\{\varphi_n\}$ is complete. Conversely, suppose that $\{\varphi_n\}$ is complete, that is, that every $f \in L_2$ can be approximated with arbitrary accuracy (in the sense of the metric in L_2) by linear combinations

$$\sum_{k=1}^{n} a_k \varphi_k$$

of elements of the set $\{\varphi_k\}$; then the partial sums

$$\sum_{k=1}^{n} c_k \varphi_k$$

of the Fourier series of f yield, in general, a still better approximation of f. Consequently, the series

$$\sum_{k=1}^{\infty} c_k \varphi_k$$

converges to f in the mean, and Parseval's equality holds.

In §53 we proved the existence of a complete orthonormal set in L_2. Inasmuch as closure and completeness are equivalent for orthonormal sets in L_2, the existence of closed orthonormal sets in L_2 need not be proved, and the examples of complete orthonormal sets in §53 are also examples of closed sets.

Bessel's inequality (7) implies that in order that numbers $c_1, c_2, \cdots$ be the Fourier coefficients of a function $f \in L_2$ with respect to an orthonormal set it is necessary that the series

$$\sum_{k=1}^{\infty} c_k^{\,2}$$

converge. In fact, this condition is not only necessary, but also sufficient. This result is stated in

THEOREM 2. (THE RIESZ-FISHER THEOREM.) *Let $\{\varphi_n\}$ be an arbitrary orthonormal set in L_2, and let the numbers*

$$c_1, \cdots, c_n, \cdots$$

be such that the series

(9)
$$\sum_{k=1}^{\infty} c_k^2$$

converges. Then there exists a function $f \in L_2$ such that

$$c_k = (f, \varphi_k),$$

and

$$\sum_{k=1}^{\infty} c_k^2 = (f, f).$$

Proof. Set

$$f_n = \sum_{k=1}^{n} c_k \varphi_k.$$

Then

$$\| f_{n+p} - f_n \|^2 = \| c_{n+1}\varphi_{n+1} + \cdots + c_{n+p}\varphi_{n+p} \|^2 = \sum_{k=n+1}^{n+p} c_k^2.$$

Since the series (9) converges, it follows, in view of the completeness of L_2, that the sequence $\{f_n\}$ converges in the mean to a function $f \in L_2$. Furthermore,

(10)
$$(f, \varphi_i) = (f_n, \varphi_i) + (f - f_n, \varphi_i),$$

where the first term on the right is equal to c_i $(n \geq i)$, and the second term approaches zero as $n \to \infty$, since

$$|(f - f_n, \varphi_i)| \leq \| f - f_n \| \cdot \| \varphi_i \|.$$

The left side of (10) is independent of n; hence, passing to the limit as $n \to \infty$, we obtain

$$(f, \varphi_i) = c_i.$$

Since, according to the definition of $f(x)$,

$$\| f - f_n \| \to 0 \qquad\qquad (n \to \infty),$$

it follows that

$$\sum_{k=1}^{\infty} c_k^2 = (f, f).$$

This proves the theorem.

In conclusion, we prove the following useful theorem:

THEOREM 3. *In order that an orthonormal set of functions* (2) *be complete it is necessary and sufficient that there not exist in L_2 a function not equivalent to $\psi \equiv 0$ which is orthogonal to all the functions of* (2).

Proof. Suppose that the set (2) is complete, and hence closed. If $f \in L_2$

is orthogonal to all the functions of (2), then all its Fourier coefficients are equal to zero. Then Parseval's equality implies that

$$(f, f) = \sum c_k^2 = 0,$$

that is, $f(x)$ is equivalent to $\psi(x) \equiv 0$.

Conversely, suppose that $\{f_n\}$ is not complete, that is, there exists a function $g \in L_2$ such that

$$(g, g) > \sum_{k=1}^{\infty} c_k^2 \qquad [c_k = (g, \varphi_k)].$$

Then, by the Riesz-Fisher theorem, there exists a function $f \in L_2$ such that

$$(f, \varphi_k) = c_k, \qquad (f, f) = \sum_{k=1}^{\infty} c_k^2.$$

The function $f - g$ is orthogonal to all the functions φ_i. In view of the inequality

$$(f, f) = \sum_{k=1}^{\infty} c_k^2 < (g, g),$$

$f - g$ cannot be equivalent to $\psi(x) \equiv 0$. This proves the theorem.

EXERCISES

1. Let $\{\varphi_n(x)\}$ be an orthonormal set in L_2 and suppose $f \in L_2$. Verify that $f - \sum_{k=1}^{n} a_k \varphi_k$ is orthogonal to all linear combinations $\sum_{k=1}^{n} b_k \varphi_k$ if, and only if, $a_k = (f, \varphi_k)$ $(1 \leq k \leq n)$.

2. Let $\{\varphi_n(x)\}$ be an orthonormal set in L_2 and let $F \subseteq L_2$ be dense in L_2. If Parseval's equality holds for each $f \in F$, then it holds for all $g \in L_2$, i.e., $\{\varphi_n(x)\}$ is closed.

This may be proved as follows: Let $s_n(f) = \sum_{k=1}^{n} c_k \varphi_k(x)$ be the nth partial sum of the Fourier series of $f \in L_2$.

a) If $f, g \in L_2$, then

$$\| s_n(f - g) \| = \| s_n(f) - s_n(g) \| \leq \| f - g \|.$$

b) Parseval's equality holds for g if, and only if,

$$\lim_{n \to \infty} \| g - s_n(g) \| = 0.$$

c) Now use the hypothesis of the exercise.

3. Let $\{\varphi_n(x)\}$, $\{\psi_n(x)\}$ be complete orthonormal sets in $L_2(R, \mu)$. Let $\mu^2 = \mu \otimes \mu$ and consider $L_2(R \times R, \mu^2)$.

a) The set $\{\chi_{nm}(x, y) = \varphi_n(x)\psi_m(y) : n, m = 1, 2, \cdots\}$ is orthonormal in $L_2(R \times R, \mu^2)$.

b) The set $\{\chi_{nm}(x, y)\}$ is complete.

Hint: Use Fubini's theorem and the criterion of Theorem 3 for completeness.

§55. Isomorphism of the spaces L_2 and l_2

The Riesz-Fisher theorem immediately implies the following important
THEOREM. *The space L_2 is isomorphic to the space l_2.*

[Two Euclidean spaces R and R' are said to be *isomorphic* if there is a
one-to-one correspondence between their elements such that

$$x \leftrightarrow x', \qquad y \leftrightarrow y'$$

implies that
  1) $x + y \leftrightarrow x' + y'$,
  2) $\alpha x \leftrightarrow \alpha x'$,
  3) $(x, y) \leftrightarrow (x', y')$.

Obviously, two isomorphic Euclidean spaces, considered merely as
metric spaces, are isometric.]

Proof. Choose an arbitrary complete orthonormal set $\{\varphi_n\}$ in L_2 and assign to each function $f \in L_2$ the sequence $c_1, \cdots, c_n, \cdots$ of its Fourier
coefficients with respect to this set. Since $\sum c_k^2 < \infty$, $(c_1, \cdots, c_n, \cdots)$
is an element of l_2. Conversely, in view of the Riesz-Fisher theorem, for
every element $(c_1, \cdots, c_n, \cdots)$ of l_2 there is an $f \in L_2$ whose Fourier
coefficients are $c_1, \cdots, c_n, \cdots$. This correspondence between the elements
of L_2 and l_2 is one-to-one. Furthermore, if

$$f^{(1)} \leftrightarrow (c_1^{(1)}, \cdots, c_n^{(1)}, \cdots)$$

and

$$f^{(2)} \leftrightarrow (c_1^{(2)}, \cdots, c_n^{(2)}, \cdots),$$

then

$$f^{(1)} + f^{(2)} \leftrightarrow (c_1^{(1)} + c_1^{(2)}, \cdots, c_n^{(1)} + c_n^{(2)}, \cdots)$$

and

$$kf^{(1)} \leftrightarrow (kc_1^{(1)}, \cdots, kc_n^{(1)}, \cdots),$$

that is, addition and multiplication by scalars are preserved by the correspondence. In view of Parseval's equality it follows that

$$(1) \qquad (f^{(1)}, f^{(2)}) = \sum_{n=1}^{\infty} c_n^{(1)} c_n^{(2)}.$$

For, the relations

$$(f^{(1)}, f^{(1)}) = \sum (c_n^{(1)})^2, \qquad (f^{(2)}, f^{(2)}) = \sum (c_n^{(2)})^2$$

and

$$\begin{aligned}
(f^{(1)} + f^{(2)}, f^{(1)} + f^{(2)}) &= (f^{(1)}, f^{(1)}) + 2(f^{(1)}, f^{(2)}) + (f^{(2)}, f^{(2)}) \\
&= \sum (c_n^{(1)} + c_n^{(2)})^2 \\
&= \sum (c_n^{(1)})^2 + 2 \sum c_n^{(1)} c_n^{(2)} + \sum (c_n^{(2)})^2
\end{aligned}$$

imply (1). Hence the above correspondence between the elements of L_2 and l_2 is an isomorphism. This proves the theorem.

On the basis of this theorem we may regard l_2 as a "coordinate form" of L_2. It enables us to carry over to L_2 results previously established for l_2. For instance, we proved in Chapter III that every linear functional in l_2 is of the form

$$\varphi(x) = (x, y),$$

where y is an element of l_2 uniquely determined by the functional φ. In view of this and the isomorphism between L_2 and l_2, it follows that every functional in L_2 is of the form

$$\varphi(f) = (f, g) = \int f(x)g(x)\, d\mu,$$

where $g(x)$ is a fixed function of L_2. We proved in §24 that $\bar{l}_2 = l_2$. Hence $\bar{L}_2 = L_2$.

The isomorphism between L_2 and l_2 established above is closely related to the theory of quantum mechanics. Quantum mechanics originally consisted of two superficially distinct theories: Heisenberg's matrix mechanics and Schrödinger's wave mechanics. Schrödinger later showed that these two theories are equivalent. From the mathematical point of view, the difference between the two theories reduced to the fact that the Heisenberg theory used the space l_2, while the Schrödinger theory used the space L_2.

EXERCISES

1. Let $\{\varphi_n(x)\}$ be an orthonormal set in L_2. Then nonequivalent functions f, g have distinct Fourier series, i.e., for some n, $(f, \varphi_n) \neq (g, \varphi_n)$ if, and only if, $\{\varphi_n\}$ is complete. This result justifies the statement in the text that $f \leftrightarrow (c_1, c_2, \cdots, c_n, \cdots)$ is a one-to-one correspondence. Hint: Apply Theorem 3 of §54.

2. Let $\{\varphi_n\}$ be a complete orthonormal set in $L_2(R, \mu)$ and suppose $f \in L_2$. The Fourier series of $f(x)$ can be integrated term by term over an arbitrary measurable subset A of R, i.e.,

$$\int_A f(x)\, d\mu = \sum_{k=1}^{\infty} c_k \int_A \varphi_k(x)\, d\mu,$$

where $c_k = (f, \varphi_k)$ is the kth Fourier coefficient of $f(x)$.

Hint: Let $f^{(1)} = f$, $f^{(2)} = \chi_A$ in equation (1) of the theorem of this section.

Chapter IX

ABSTRACT HILBERT SPACE. INTEGRAL EQUATIONS WITH SYMMETRIC KERNEL

In the preceding chapter we proved that a separable L_2 is isomorphic with l_2, i.e., that they are two essentially different realizations of the same space. This space, usually called Hilbert space, plays an important part in analysis and its applications. It is often convenient not to restrict oneself, as previously, to various realizations of Hilbert space, but to define it axiomatically, for instance, as Euclidean n-space is defined in linear algebra.

§56. Abstract Hilbert space

DEFINITION 1. A set H of arbitrary elements $f, g, \cdots, h, \cdots$ is called an (abstract) Hilbert space if:

I. H is a linear space.

II. An inner product is defined in H, i.e., every pair of elements f, g is assigned a real number (f, g) such that

1) $(f, g) = (g, f)$,
2) $(\alpha f, g) = \alpha(f, g)$,
3) $(f_1 + f_2, g) = (f_1, g) + (f_2, g)$,
4) $(f, f) > 0$ if $f \neq 0$.

In other words, Conditions I and II mean that H is a Euclidean space. The number $\| f \| = (f, f)^{\frac{1}{2}}$ is called the *norm* of f.

III. The space H is complete in the metric $\rho(f, g) = \| f - g \|$.

IV. H is infinite-dimensional, that is, for every natural number n, H contains n linearly independent vectors.

V. H is separable. (This condition is often omitted; H may then be nonseparable.) Then H contains a countable dense set.

It is easy to give examples of spaces satisfying all the axioms. One such is the space l_2 discussed in Chapter II. In fact, l_2 is an infinite-dimensional Euclidean space, since the elements

$$e_1 = (1, 0, 0, \cdots, 0, \cdots)$$
$$e_2 = (0, 1, 0, \cdots, 0, \cdots)$$
$$e_3 = (0, 0, 1, \cdots, 0, \cdots)$$
$$\cdots\cdots\cdots\cdots\cdots\cdots\cdots\cdots$$

are linearly independent; it was proved in §§9 and 13 of Chapter II that it is complete and separable. The space L_2 of functions square integrable with respect to a separable measure, which is isomorphic to l_2, also satisfies the same axioms.

The following proposition holds:

All Hilbert spaces are isomorphic.

To prove this, it is obviously sufficient to show that every Hilbert space is isomorphic to the coordinate space l_2. The latter assertion is proved by essentially the same arguments as were used in the proof of the isomorphism of L_2 and l_2 :

1. The definitions of orthogonality, closure and completeness, which were introduced in §53 for elements of L_2, can be transferred unchanged to abstract Hilbert space.

2. Choosing in H a countable dense set and applying to it the process of orthogonalization described (for L_2) in §53, we construct in H a complete orthonormal set, that is, a set

$$(1) \qquad\qquad h_1, \cdots, h_n, \cdots$$

satisfying:

a)

$$(h_i, h_k) = \begin{cases} 0 & (i \neq k), \\ 1 & (i = k). \end{cases}$$

b) The linear combinations of the elements of (1) are dense in H.

3. Let f be an arbitrary element of H. Set $c_k = (f, h_k)$. Then the series $\sum c_k^2$ converges, and $\sum c_k^2 = (f, f)$ for an arbitrary complete orthonormal set $\{h_k\}$ and $f \in H$.

4. Suppose again that $\{h_k\}$ is a complete orthonormal set in H. If

$$c_1, \cdots, c_n, \cdots$$

is a sequence of numbers such that

$$\sum c_k^2 < \infty,$$

there exists an $f \in H$ such that

$$c_k = (f, h_k),$$

and

$$\sum c_k^2 = (f, f).$$

5. It is clear from what we have said that an isomorphism between H and l_2 can be realized by setting

$$f \leftrightarrow (c_1, \cdots, c_n, \cdots),$$

where

$$c_k = (f, h_k)$$

and

$$h_1, h_2, \cdots, h_n, \cdots$$

is an arbitrary complete orthonormal set in H.

The reader may carry out the details of the proof as in §§53–55.

EXERCISES

1. a) The norm $\| f \|$ in H satisfies the parallelogram law:

$$\| f_1 + f_2 \|^2 + \| f_1 - f_2 \|^2 = 2(\| f_1 \|^2 + \| f_2 \|^2).$$

b) Conversely, if X is a complete separable normed linear space in which the norm satisfies the parallelogram law, then an inner product may be defined in X by

$$(f, g) = \tfrac{1}{4}[\| f + g \|^2 - \| f - g \|^2].$$

Moreover, $(f, f) = \| f \|^2$ and X becomes a Hilbert space. Hints: (i) Establish first that (x, y) is a continuous function of x. (ii) Show by induction that $(nx, y) = n(x, y)$. It will then readily follow that $(ax, y) = a(x, y)$. (iii) Then establish that $(x_1 + x_2, y) = (x_1, y) + (x_2, y)$. The other properties of an inner product are immediate.

2. Suppose that $A \subseteq H$ has the property that $f, g \in A$ implies that $\tfrac{1}{2}(f + g) \in A$ (this is true, in particular, if A is convex). Let

$$d = \inf \{ \| f \| : f \in A \}.$$

If $\{f_n\} \subset A$ has the property that $\lim_{n \to \infty} \| f_n \| = d$, show that $\{f_n\}$ is a fundamental sequence in H.

Since H is complete, it follows that $\lim_{n \to \infty} f_n = f$ exists in H. If A is closed, then $f \in A$.

Hint: The parallelogram law yields

$$\| \tfrac{1}{2}(f_n - f_m) \|^2 = \tfrac{1}{2} \| f_n \|^2 + \tfrac{1}{2} \| f_m \|^2 - \| \tfrac{1}{2}(f_n + f_m) \|^2$$
$$\leq \tfrac{1}{2} \| f_n \|^2 + \tfrac{1}{2} \| f_m \|^2 - d^2.$$

3. In Def. 2 of §50 it is stated that the Schwarz inequality:

$$(f, g)^2 \leq \| f \|^2 \| g \|^2$$

holds. The author proves Schwarz's inequality in several concrete cases (see vol. 1, pp. 17, 18). Prove that the inequality holds in H (only Axioms I and II of §56 are required).

Hint: Suppose that $f, g \in H$ and that t is real. The quadratic polynomial with real coefficients in t: $(f + t(f, g)g, f + t(f, g)g)$ is nonnegative; hence, its discriminant must be nonpositive.

4. The inner product (f, g) is a continuous function of f and g, i.e., if $\| f_n - f \| \to 0$ and $\| g_n - g \| \to 0$ for $\{f_n\}, \{g_n\}, f, g$ in H, then $(f_n, g_n) \to (f, g)$.

5. The following is an example of a nonseparable Hilbert space. Let H be the collection of all real-valued functions defined on $[0, 1]$ with $f(x) \neq 0$ for only countably many $x \in [0, 1]$, and such that if $f(x_n) \neq 0$ for $\{x_n\}$, then $\sum [f(x_n)]^2 < \infty$. The addition of functions and multiplication by real numbers is defined in the usual way, i.e., pointwise.

a) Define (f, g) analogously to the scalar product in l_2 and show that H satisfies I, II, III, IV in Def. 1.

b) Show that H is not separable. Hint: Show that H contains uncountably many disjoint open spheres.

6. If in the preceding example we restrict the collection H further to those $f(x)$ whose values are not zero for only finitely many x, then with the same operations H is also an incomplete metric space. That is, find a fundamental sequence in H whose limit is not in H. Note that the limit of such a sequence will be an element of H in Ex. 5.

§57. Subspaces. Orthogonal complements. Direct sums

In accordance with the general definitions of Chapter III, §21, a *linear manifold* in H is a subset L of H such that if $f, g \in L$, then $\alpha f + \beta g \in L$ for arbitrary numbers α and β. A *subspace* of H is a closed linear manifold in H.

We give several examples of subspaces of H.

1. Suppose that $h \in H$ is arbitrary. The set of all $f \in H$ orthogonal to h is a subspace of H.

2. Let $H = l_2$, that is, all the elements of H are sequences

$$(x_1, \cdots, x_n, \cdots)$$

of numbers such that $\sum x_k^2 < \infty$. The elements satisfying the condition $x_1 = x_2$ form a subspace.

3. Let H be the space L_2 of all square summable functions on a closed interval $[a, b]$ and suppose that $a < c < b$. We denote by H_c the collection of all functions of H identically zero on $[a, c]$. H_c is a subspace of H. If $c_1 < c_2$, then $H_{c_1} \supset H_{c_2}$, and $H_a = H$, $H_b = (0)$. Hence we obtain a continuum of subspaces of H ordered by inclusion. Each of these subspaces (with the exception, of course, of H_b) is infinite-dimensional and isomorphic to H.

The verification of the fact that each of the sets described in 1–3 is indeed a subspace of H is left to the reader.

Every subspace of a Hilbert space is either a finite-dimensional Euclidean space or itself a Hilbert space. For, Axioms I–III are obviously satisfied by a subspace and the validity of Axiom V follows from the following lemma:

LEMMA. *If a metric space R contains a countable dense set, every subspace R' of R contains a countable dense set.*

Proof. [TRANS. NOTE. The proof in the original was incorrect. We have therefore substituted the following proof.]

We assume that $R' \neq \emptyset$, otherwise there is nothing to prove.

Let $\{\xi_n\}$ be dense in R. For every pair of natural numbers n, k choose a $\zeta_{nk} \in R'$ (if it exists) such that

$$\rho(\xi_n, \zeta_{nk}) < 1/2k.$$

Then $\{\zeta_{nk}; n, k = 1, 2, \cdots\}$ is dense in R'. To see this, suppose that $x \in R'$ and $\epsilon > 0$. Choose an s such that $1/s < \epsilon$. Since $\{\xi_n\}$ is dense in R and $x \in R$, there is an m such that $\rho(\xi_m, x) < 1/2s$. Hence ζ_{ms} exists and

$$\rho(x, \zeta_{ms}) \leq \rho(x, \xi_m) + \rho(\xi_m, \zeta_{ms}) < 1/2s + 1/2s = 1/s < \epsilon.$$

The existence in Hilbert space of an inner product and the notion of orthogonality enable us to supplement substantially the results of Vol. 1 on subspaces of arbitrary Banach spaces.

By orthogonalizing a countable dense sequence of elements of an arbitrary subspace of a Hilbert space, we obtain

THEOREM 1. *Every subspace M of H contains an orthogonal set $\{\varphi_n\}$ whose linear closure coincides with M:*

$$M = \bar{M}(\varphi_1, \cdots, \varphi_n, \cdots).$$

Let M be a subspace of H. Denote by

$$M' = H \ominus M$$

the set of $g \in H$ orthogonal to all $f \in M$. We shall prove that M' is also a subspace of H. The linearity of M' is obvious, since $(g_1, f) = (g_2, f) = 0$ implies that $(\alpha_1 g_1 + \alpha_2 g_2, f) = 0$. To prove closure, suppose that $g_n \in M'$ and that g_n converges to g. Then

$$(g, f) = \lim_{n \to \infty} (g_n, f) = 0$$

for all $f \in M$, and consequently $g \in M'$.

M' is called the *orthogonal complement* of M.

From Theorem 1 it easily follows that:

THEOREM 2. *If M is a subspace of H, every $f \in H$ is uniquely representable in the form $f = h + h'$, where $h \in M$, $h' \in M'$.*

Proof. We shall first prove the existence of the decomposition. To this end, we choose in M a complete orthonormal set $\{\varphi_n\}$ such that $M = \bar{M}\{\varphi_n\}$ and set

$$h = \sum_{n=1}^{\infty} c_n \varphi_n, \qquad c_n = (f, \varphi_n).$$

Since $\sum c_n^2$ converges (by the Bessel inequality), h exists and is an element of M. Set

$$h' = f - h.$$

Obviously,

$$(h', \varphi_n) = 0$$

for all n. Inasmuch as an arbitrary element ζ of M can be written as

$$\zeta = \sum a_n \varphi_n,$$

we have

$$(h', \zeta) = \sum_{n=1}^{\infty} a_n (h', \varphi_n) = 0.$$

We now suppose that, in addition to the above decomposition $f = h + h'$, there is another one:

$$f = h_1 + h_1', \qquad (h_1 \in M, h_1' \in M').$$

Then

$$(h_1, \varphi_n) = (f, \varphi_n) = c_n.$$

It follows that

$$h_1 = h, \qquad h_1' = h'.$$

Theorem 2 implies

COROLLARY 1. *The orthogonal complement of the orthogonal complement of a subspace M coincides with M.*

It is thus possible to speak of complementary subspaces of H. If M and M' are two complementary subspaces and $\{\varphi_n\}$, $\{\varphi_n'\}$ are complete orthonormal sets in M and M', respectively, the union of the sets $\{\varphi_n\}$ and $\{\varphi_n'\}$ is a complete orthonormal set in H. Therefore,

COROLLARY 2. *Every orthonormal set $\{\varphi_n\}$ can be extended to a set complete in H.*

If the set $\{\varphi_n\}$ is finite, the number of its terms is the dimension of M and also the deficiency of M'. Hence

COROLLARY 3. *The orthogonal complement of a subspace of finite dimension n has deficiency n, and conversely.*

If every vector $f \in H$ is represented in the form $f = h + h'$, $h \in M$, $h' \in M'$ (M' the orthogonal complement of M), we say that H is the *direct sum* of the orthogonal subspaces M and M' and write

$$H = M \oplus M'.$$

It is clear that the notion of a direct sum can be immediately generalized to an arbitrary finite or even countable number of subspaces: H is the direct sum of subspaces $M_1, \cdots, M_n, \cdots$:

$$H = M_1 \oplus \cdots \oplus M_n \oplus \cdots$$

if

1) the subspaces M_i are pairwise orthogonal, that is, an arbitrary vector in M_i is orthogonal to an arbitrary vector in M_k ($i \neq k$);

2) every $f \in H$ can be written in the form

$$(1) \qquad\qquad f = h_1 + \cdots + h_n + \cdots \qquad\qquad (h_n \in M_n),$$

where $\sum \| h_n \|^2$ converges if the number of subspaces M_n is infinite.

It is easily verified that the sum (1) is unique and that

$$\| f \|^2 = \sum_n \| h_n \|^2.$$

A notion related to the direct sum of subspaces is that of the direct sum of a finite or countable number of arbitrary Hilbert spaces. If H_1, H_2 are Hilbert spaces, their direct sum H is defined as follows: the elements of H are all possible pairs (h_1, h_2), where $h_1 \in H_1$, $h_2 \in H_2$, and the inner product of two such pairs is

$$((h_1 ; h_2), (h_1', h_2')) = (h_1, h_1') + (h_2, h_2').$$

The space H obviously contains the orthogonal subspaces consisting of pairs of the form $(h_1, 0)$ and $(0, h_2)$, respectively; the first can be identified in a natural way with the space H_1, and the second with H_2.

The sum of an arbitrary finite number of spaces is defined in the same way. The sum $H = \sum_n \oplus H_n$ of a countable number of spaces $H_1, \cdots, H_n, \cdots$ is defined as follows: the elements of H are all possible sequences of the form

$$h = (h_1, \cdots, h_n, \cdots)$$

such that

$$\sum_n \| h_n \|^2 < \infty.$$

The inner product (h, g) of $h, g \in H$ is equal to

$$\sum_n (h_n, g_n).$$

EXERCISES

1. Prove Corollary 1 of Theorem 2.

2. Prove the remark before Corollary 2: If M, M' are complementary subspaces and if $\{\varphi_n\}$, $\{\varphi_n'\}$ are complete orthonormal sets in M, M', respectively, then their union is a complete orthonormal set in H.

3. If M, N are orthogonal subspaces, then

$$M + N = \{f + g : f \in M, g \in N\}$$

is closed, and therefore a subspace. If M, N are not orthogonal, the result need not be true. (An example can be found in Halmos, P. R., *Introduction to Hilbert Space and the Theory of Spectral Multiplicity*, New York, 1951.)

Hint: It is enough to show that if $\{f_n + g_n\}$ is a fundamental sequence, then $\{f_n\}$ and $\{g_n\}$ are also fundamental.

It is the purpose of the following exercises to extend some of the results of the text to general (i.e., nonseparable) Hilbert spaces.

4. If a subspace M of H is proper, i.e., $H \setminus M \neq \emptyset$, then there exists an element g of H, $g \neq 0$, such that g is orthogonal to every element of M.

Hint: For $h \in H \setminus M$, $h - M = \{h - x : x \in M\}$ is closed and convex. Let $d = \inf \{\| h - x \| : x \in M\}$, $d = \| h - x_0 \|$, $x_0 \in M$ (see §56, Ex. 2). For arbitrary real c and $f \in M$, show that

$$0 \leq \| h - (x_0 + cf) \|^2 - \| h - x_0 \|^2.$$

Show that this holds only if $g = h - x_0$ is orthogonal to M. g will be the required element.

5. If M, N are subspaces of H, $N \subseteq M$, then we denote by $M \ominus N$ the orthogonal complement of N in M (consider M itself as a Hilbert space). Show that $M = N \oplus (M \ominus N)$.

Hint: Let $L = N \oplus (M \ominus N)$, with $L \subseteq M$ and L closed (see Ex. 3). If L is properly contained in M, apply the result of Ex. 4 to obtain a contradiction.

6. Let $F(f)$ be a bounded linear functional on H. There exists one and only one element g in H such that $F(f) = (f, g)$ for every f in H [compare with equation (1) at the beginning of the next section].

   a) The uniqueness is easy to establish.

   b) Let $M = \{f : F(f) = 0\}$. M is a subspace. If $M = H$, choose $g = 0$. Otherwise, by Ex. 4 there exists an $h \neq 0$ such that h is orthogonal to M. Show that $g = [F(h)/(h, h)]h$ will do.

§58. Linear and bilinear functionals in Hilbert space

The isomorphism of every Hilbert space with l_2 enables us to carry over to an abstract Hilbert space the results established in Chapter III for l_2.

Since every linear functional in l_2 is of the form

$$\varphi(x) = (x, a) \qquad (a \in l_2),$$

it follows that:

An arbitrary linear functional $F(h)$ in H is of the form

(1) $$F(h) = (h, g),$$

where g depends only on F.

Hence, the definition of weak convergence introduced in Chapter III for an arbitrary linear space, when applied to H can be stated in the following way:

A sequence $h_n \in H$ is weakly convergent to $h \in H$ if

1) the norms $\| h_n \|$ are bounded (see p. 90 of vol. 1);
2) for every $g \in H$,

$$(h_n, g) \to (h, g).$$

Ex. 4 at the end of the section shows that 2) implies 1).

An arbitrary orthonormal sequence

$$\varphi_1, \cdots, \varphi_n, \cdots$$

in H converges weakly to zero, since

$$c_n = (h, \varphi_n) \to 0 \qquad\qquad (n \to \infty)$$

for arbitrary $h \in H$, in view of the fact that

$$\sum c_n^2 \leq (h, h) < \infty.$$

Such a sequence, of course, does not converge in the norm.

In particular, applying these remarks to the case when H is the space of square integrable functions on a closed interval $[a, b]$ of the real line with the usual Lebesgue measure, we obtain the following interesting result: Let

$$\varphi_1(t), \cdots, \varphi_n(t), \cdots$$

be an orthonormal set of functions in H, and let

$$f(t) = \begin{cases} 1 & (\text{on } [t_1, t_2] \subset [a, b]), \\ 0 & (\text{outside } [t_1, t_2]). \end{cases}$$

Then

$$(f, \varphi_n) = \int_{t_1}^{t_2} \varphi_n(t) \, dt.$$

Hence

$$\int_{t_1}^{t_2} \varphi_n(t) \, dt \to 0$$

for an arbitrary orthonormal set of functions $\varphi_n(t)$ and arbitrary t_1, $t_2 \in [a, b]$.

If the $\varphi_n(t)$ are uniformly bounded,

$$\int_a^b \varphi_n^2(t) \, dt = 1$$

only if the number of sign changes of $\varphi_n(t)$ on $[a, b]$ is unbounded as $n \to \infty$ (the same is to be observed, for instance, in the case of trigonometric functions).

In Chapter III, parallel with the concept of weak convergence of the elements of a linear normed space, we introduced the notion of weak convergence of a sequence of functionals. Inasmuch as Hilbert space coincides with its conjugate space, these two types of convergence are identical. Therefore, Theorem $1'$ of §28 yields the following result for Hilbert space H:

The unit sphere in H is weakly compact, that is, every sequence $\varphi_n \in H$, with $\| \varphi_n \| \leq 1$, contains a weakly convergent subsequence.

In the sequel we require in addition the following

THEOREM 1. *If ξ_n is weakly convergent to ξ in H, then*

$$\| \xi \| \leq \sup \| \xi_n \|.$$

Proof. For every complete orthonormal set $\{\varphi_k\}$ in H,

$$c_k = (\xi, \varphi_k) = \lim_{n \to \infty} (\xi_n, \varphi_k) = \lim_{n \to \infty} c_{nk},$$

$$\sum_{m=1}^{k} c_m^2 = \lim_{n \to \infty} \sum_{m=1}^{k} c_{nm}^2 \leq \sup_n \sum_{m=1}^{\infty} c_{nm}^2;$$

consequently,

$$\sum_{m=1}^{\infty} c_m^2 \leq \sup_n \sum_{m=1}^{\infty} c_{nm}^2,$$

which proves the theorem.

Let $B(f, g)$ be a real-valued function of pairs of elements of H satisfying the following condition: $B(f, g)$ is a linear functional of f for fixed g, and a linear functional of g for fixed f. $B(f, g)$ is called a *bilinear functional*. A bilinear functional $B(f, g)$ is said to be *symmetric* if

$$B(f, g) = B(g, f) \qquad\qquad (f, g \in H)$$

The theorem on the general form of a linear functional in H implies that every bilinear functional in H can be written in the form

$$B(f, g) = (\zeta, g),$$

where ζ depends on f. It is easily seen that the correspondence

$$f \to \zeta$$

is a continuous linear operator in H; denote it by A. Hence

(2) $$B(f, g) = (Af, g).$$

An alternative form

$$B(f, g) = (f, A^*g),$$

where A^* is the adjoint operator of A, can be obtained in a similar fashion. [In Chapter III, in considering linear operators on an arbitrary Banach space E, we defined the adjoint operator A^* of A by means of the relation

$$(Ax, \varphi) = (x, A^*\varphi) \qquad (x \in E, \varphi \in \bar{E}).$$

If E is a Hilbert space, then $\bar{E} = E$, and the definition of A^* in Chapter III reduces to the definition given above.] If a functional $B(f, g)$ is symmetric, then

$$(Af, g) = B(f, g) = B(g, f) = (Ag, f) = (f, Ag),$$

that is,

(3) $$A = A^*.$$

A linear operator satisfying (3) is said to be *self-adjoint*.

Formula (2) defines a one-to-one correspondence between the bilinear functionals and the continuous linear operators on H, with the symmetric bilinear functionals corresponding to the self-adjoint linear operators, and conversely.

Setting $f = g$ in a symmetric bilinear functional, we obtain a *quadratic functional*

$$Q(f) = B(f, f).$$

According to (2),

$$Q(f) = (Af, f),$$

where A is a self-adjoint linear operator.

Since the correspondence between the symmetric bilinear functionals and the quadratic functionals is one-to-one $[Q(f) = B(f, f)$, and conversely: $B(f, g) = \frac{1}{4}\{Q(f + g) - Q(f - g)\}]$, the correspondence between the quadratic functionals and the self-adjoint linear operators is also one-to-one.

EXERCISES

1. Let M be a proper subspace of H, $F(h)$ a (bounded) linear functional on M with norm $\| F \|$. Then there exists a linear functional F^* on H such that $F^*(h) = F(h)$ for $h \in M$ and $\| F^* \| = \| F \|$. (See vol. 1, p. 86, the Hahn-Banach theorem.) Hint: Apply equation (1) at the beginning of this section to the Hilbert space M.

2. a) Let $\{f_n\}$ and $f(x)$ belong to H. If $\{f_n\}$ converges strongly to f, then $\{f_n\}$ converges weakly to f.

b) The converse is false. Show that for $L_2(R, \mu)$, where $R = [0, 1]$ and μ is linear Lebesgue measure, that $\{\sin nx\}$ is weakly convergent to

$f(x) \equiv 0$, but $\{\sin nx\}$ is not fundamental in the norm; hence it cannot converge strongly to any element of L_2.

c) The following partial converse is true. If $\{f_n\}$ converges weakly to f and $\| f_n \|$ converges to $\| f \|$, then $\{f_n\}$ converges strongly to f. Hint: Show that $(f - f_n, f - f_n) \to 0$.

3. Suppose $A \subseteq H$. If A is weakly closed, then A is a norm closed subset of H. More explicitly: Suppose $\{f_n\} \subset A$ and $(f_n, g) \to (f, g)$ for every $g \in H$ implies that $f \in A$. Show that $\| f_n - f \| \to 0$ and $\{f_n\} \subset A$ implies $f \in A$.

4. The definition of weak convergence of $\{h_n\}$ to h in H lists two conditions. We propose to show that the second condition already implies the first. This result in a more general setting is known as the Banach-Steinhaus theorem.

a) It is enough to show that there exists a constant $M > 0$ and a sphere $S = \{g \colon \| g - g_0 \| \leq r\}$ such that $g \in S$ implies $| (h_n, g) | \leq M$. For if this implication holds and $\| g \| \leq r$, then

$$| (h_n, g) | = | (h_n, g + g_0) - (h_n, g_0) | \leq 2M.$$

Now show that $g \in H$ implies

$$| (h_n, g) | \leq (2M/r) \| g \|$$

and consequently

$$\| h_n \| \leq 2M/r.$$

b) It follows that if the result is false, the sequence $\{| (h_n, g) |\}$ must be unbounded in every sphere, i.e., given $a > 0$ and S a sphere in H, there is an element $g_a \in S$ and an index n_a for which $| (h_{n_a}, g_a) | > a$. Show by continuity of (h, g) in the second argument that S contains a closed sphere S_a such that $g \in S_a$ implies $| (h_{n_a}, g) | > a$.

c) Now construct by induction a sequence of closed spheres $\{S_k\}$ and a sequence $\{n_k\}$ such that $S_k \subseteq S_{k-1}$; diam $S_k \leq 1/k; n_1 < n_2 < \cdots < n_k < \cdots ; n_k \to \infty$ and $| (h_{n_k}, g) | > k$ for $g \in S_k$.

d) Use the completeness of the metric space H to show the existence of a point g_0 for which $| (h_{n_k}, g_0) | > k$. This contradicts Condition 2): $(h_n, g_0) \to (h, g_0)$.

5. Let M_1 and M_2 be two subspaces with M_1 a proper subset of M_2. Show that there exists an element g of M_2 such that $\| g \| = 1$ and $\| g - f \| \geq 1$ for any $f \in M_1$. Hint: Consider M_2 as a Hilbert space with subspace M_1.

6. a) Let X be a Banach space, M a subspace of X and $\bar{x} = M + x = \{y + x \colon y \in M\}$ a subset of X defined for each $x \in X$. We can make the collection $\{\bar{x} \colon x \in X\} = X/M$ into a vector space, called the *quotient space*

of X mod M by defining $\bar{x} + \bar{y} = \{M + (x + y)\}$ and $a\bar{x} = \{M + ax\}$. Verify that the operations are uniquely defined and that X/M is a vector space with zero element $\bar{0} = M$.

b) A norm is introduced by defining

$$\| \bar{x} \| = \inf \{\| y + x \| : y \in M\}.$$

Show that X/M is a normed linear space. The fact that M is closed is required to show that $\| \bar{x} \| = 0$ if, and only if, $\bar{x} = \bar{0}$.

c) It is also true that X/M is complete, i.e., X/M is a Banach space. This is more difficult to prove.

7. If $X = H$ and M is a subspace of H, show that H/M and $H \ominus M$ are isomorphic normed vector spaces. In other words, suppose that $f \in H$, $f = h + h'$, with $h \in M$, $h' \in M'$. Suppose that $\bar{f}$ corresponds to h' and show that this correspondence is one-to-one, onto and preserves addition, multiplication by scalars, and the norm. Hint: $\| f \|^2 = \| h \|^2 + \| h' \|^2$.

§59. Completely continuous self-adjoint operators in H

In Chapter IV we introduced the notion of a completely continuous linear operator, acting on a Banach space E. In this section we restrict the discussion to self-adjoint completely continuous operators acting on a Hilbert space, supplemented by the results already established for arbitrary completely continuous operators.

We recall that we called an operator A completely continuous if it mapped every bounded set into a compact set. Inasmuch as $H = \bar{H}$, that is, H is conjugate to a separable space, the bounded sets in H are precisely the weakly compact sets (see Ex. 1 at the end of the section). Therefore, the definition of a completely continuous operator on a Hilbert space can be stated as follows:

An operator A acting on a Hilbert space H is said to be completely continuous if it maps every weakly compact set into a compact set (relative to the norm).

In a Hilbert space this is equivalent to the condition that the operator A map every weakly convergent sequence into a norm convergent sequence (see Ex. 2 at the end of the section).

In this section we shall prove the following fundamental theorem, a generalization to completely continuous operators of the theorem on the reduction of the matrix of a self-adjoint linear transformation in n-dimensional space to diagonal form:

THEOREM 1. *For every completely continuous self-adjoint linear operator A on a Hilbert space H there exists an orthonormal set of eigenvectors (characteristic vectors; see vol. 1, p. 110) $\{\varphi_n\}$, corresponding to eigenvalues (characteristic* values) $\{\lambda_n\}$, such that every $\xi \in H$ can be written uniquely in the form $\xi = \sum_k c_k \varphi_k + \xi'$, where the vector ξ' satisfies the condition $A\xi' = 0$. Also

$$A\xi = \sum_k \lambda_k c_k \varphi_k ,$$

and $\lim_{n \to \infty} \lambda_n = 0$.

For the proof of this fundamental theorem we require the following lemmas:

LEMMA 1. If $\{\xi_n\}$ converges weakly to ξ and the self-adjoint linear operator A is completely continuous, then

$$Q(\xi_n) = (A\xi_n , \xi_n) \to (A\xi, \xi) = Q(\xi).$$

Proof.

$$| (A\xi_n , \xi_n) - (A\xi, \xi) | \leq | (A\xi_n , \xi_n) - (A\xi_n , \xi) + (A\xi_n , \xi) - (A\xi, \xi) |.$$

But

$$| (A\xi_n , \xi_n) - (A\xi_n , \xi) | = | (\xi_n , A(\xi_n - \xi)) | \leq \| \xi_n \| \cdot \| A(\xi_n - \xi) \|$$

and

$$| (A\xi_n , \xi) - (A\xi, \xi) | = | (\xi, A(\xi_n - \xi)) | \leq \| \xi \| \cdot \| A(\xi_n - \xi) \|.$$

Since the numbers $\| \xi_n \|$ are bounded and $\| A(\xi_n - \xi) \| \to 0$,

$$| (A\xi_n , \xi_n) - (A\xi, \xi) | \to 0.$$

This proves the lemma.

LEMMA 2. If a functional

$$| Q(\xi) | = | (A\xi, \xi) |,$$

where A is a bounded self-adjoint linear operator, assumes a maximum at a point ξ_0 of the unit sphere, then

$$(\xi_0 , \eta) = 0$$

implies that

$$(A\xi_0 , \eta) = (\xi_0 , A\eta) = 0.$$

Proof. Obviously, $\| \xi_0 \| = 1$. Set

$$\xi = (\xi_0 + a\eta)/(1 + a^2 \| \eta \|^2)^{\frac{1}{2}},$$

where a is an arbitrary number. From $\| \xi_0 \| = 1$ it follows that

$$\| \xi \| = 1.$$

Since

$$Q(\xi) = (1 + a^2 \| \eta \|^2)^{-1}[Q(\xi_0) + 2a(A\xi_0 , \eta) + a^2 Q(\eta)],$$

it follows that

$$Q(\xi) = Q(\xi_0) + 2a(A\xi_0, \eta) + O(a^2)$$

for small values of a. It is clear from the last relation that if $(A\xi_0, \eta) \neq 0$, then a can be chosen so that $|Q(\xi)| > |Q(\xi_0)|$. This contradicts the hypothesis of the lemma.

It follows immediately from Lemma 2 that if $|Q(\xi)|$ assumes a maximum at $\xi = \xi_0$, then ξ_0 is an eigenvector of the operator A.

Proof of the theorem. We shall construct the elements φ_k by induction, in the order of decreasing absolute values of the corresponding eigenvalues:

$$|\lambda_1| \geq \cdots \geq |\lambda_n| \geq \cdots.$$

To construct the element φ_1 we consider the expression $Q(\xi) = |(A\xi, \xi)|$ and show that it assumes a maximum on the unit sphere. Let

$$S = \sup \{|(A\xi, \xi)|; \|\xi\| \leq 1\}$$

and suppose that $\xi_1, \xi_2, \cdots$ is a sequence such that $\|\xi_n\| \leq 1$ and

$$|(A\xi_n, \xi_n)| \to S \qquad (n \to \infty).$$

Since the unit sphere in H is weakly compact, $\{\xi_n\}$ contains a subsequence weakly convergent to an element η. In view of Theorem 1, §58, $\|\eta\| \leq 1$, and by Lemma 1,

$$|(A\eta, \eta)| = S.$$

We take η as φ_1. Clearly, $\|\eta\| = 1$. Also

$$A\varphi_1 = \lambda_1\varphi_1,$$

whence

$$|\lambda_1| = |(A\varphi_1, \varphi_1)|/(\varphi_1, \varphi_1) = |(A\varphi_1, \varphi_1)| = S.$$

Now suppose that the eigenvectors

$$\varphi_1, \cdots, \varphi_n$$

corresponding to the eigenvalues

$$\lambda_1, \cdots, \lambda_n$$

have already been constructed. We consider the functional

$$|(A\xi, \xi)|$$

on the elements of

$$M_n' = H \ominus M(\varphi_1, \cdots, \varphi_n)$$

(that is, the set orthogonal to $\varphi_1, \cdots, \varphi_n$) and such that $\|\xi\| \leq 1$. M_n' is an invariant subspace (a subspace which is mapped into itself) of A [since $M(\varphi_1, \cdots, \varphi_n)$ is invariant and A is self-adjoint]. Applying the above arguments to M_n, we obtain an eigenvector φ_{n+1} of A in M_n'.

Two cases are possible: 1) after a finite number of steps we obtain a subspace M_{n_0}' in which $(A\xi, \xi) \equiv 0$; 2) $(A\xi, \xi) \neq 0$ on M_n' for all n.

In the first case Lemma 2 implies that A maps M_{n_0}' into zero, that is, M_{n_0}' consists of the eigenvectors corresponding to $\lambda = 0$. The set of vectors $\{\varphi_n\}$ is finite.

In the second case we obtain a sequence $\{\varphi_n\}$ of eigenvectors for each of which $\lambda_n \neq 0$. We show that $\lambda_n \to 0$. The sequence $\{\varphi_n\}$ (like every orthonormal sequence) is weakly convergent to zero. Therefore, $A\varphi_n = \lambda_n\varphi_n$ converge to zero in the norm, whence $|\lambda_n| = \|A\varphi_n\| \to 0$.

Let

$$M' = H \ominus M\{\varphi_n\} = \cap_n M_n' \neq 0.$$

If $\xi \in M'$ and $\xi \neq 0$, then

$$(A\xi, \xi) \leq |\lambda_n| \|\xi\|^2$$

for all n, that is,

$$(A\xi, \xi) = 0.$$

Hence, applying Lemma 2 (for sup $\{|(A\xi, \xi)|; \|\xi\| \leq 1\} = 0$) to M', we obtain $A\xi = 0$, that is, A maps the subspace M' into zero.

From the construction of the set $\{\varphi_n\}$ it is clear that every vector can be written in the form

$$\xi = \sum_k c_k\varphi_k + \xi' \qquad (A\xi' = 0).$$

Hence

$$A\xi = \sum_k \lambda_k c_k\varphi_k.$$

EXERCISES

1. Let A be a continuous linear operator of H into H. Suppose that $\{f_n\} \subset H, f \in H$ and $\{f_n\}$ converges weakly to f. Show that $\{Af_n\}$ converges weakly to Af.

2. In the second paragraph of this section it is stated that in H the norm bounded sets are precisely the weakly compact sets. Show that this is true as follows:

a) If $A \subseteq H$ is norm bounded, i.e., there exists an $M > 0$ such that $\|f\| < M$ for all $f \in A$, then Theorem 1' of §28 (see vol. 1) shows that A is weakly compact (see the statement preceding Theorem 1 in §58).

b) If A is weakly compact, show that A is norm bounded. Use Ex. 4 of §58 for this purpose.

3. In the fourth paragraph of this section it is stated that the following two properties of an operator A on H are equivalent:

a) A maps every weakly compact set into a norm compact set.

b) A maps every weakly convergent sequence into a norm convergent sequence.

Prove that a) and b) are equivalent.

4. Let A be a continuous (bounded) linear operator of H into H with the additional property that $A(H)$, the range of A, is contained in a finite-dimensional subspace of H. Then A is completely continuous. Hint: The Bolzano-Weierstrass theorem holds in E_n.

5. Let A be a completely continuous operator, $T = I - A$ and suppose $M \subseteq H$, $M = \{x : Tx = 0\}$. Show that M is a finite-dimensional subspace of H.

§60. Linear equations in completely continuous operators

We consider the equation

$$(1) \qquad\qquad \xi = cA\xi + \eta,$$

where A is a completely continuous self-adjoint operator, $\eta \in H$ is prescribed and $\xi \in H$ is the unknown.

Let

$$\varphi_1, \cdots, \varphi_n, \cdots$$

be the eigenvectors of A corresponding to the eigenvalues different from zero. Then η can be written as

$$(2) \qquad\qquad \eta = \sum_n a_n \varphi_n + \eta',$$

where $A\eta' = 0$. We shall seek a solution of (1) of the form

$$(3) \qquad\qquad \xi = \sum_n x_n \varphi_n + \xi',$$

where $A\xi' = 0$. Substitution of (2) and (3) into (1) yields

$$\sum_n x_n(1 - \lambda_n c)\varphi_n + \xi' = \sum_n a_n \varphi_n + \eta'.$$

This equation is satisfied if, and only if,

$$\xi' = \eta',$$
$$x_n(1 - \lambda_n c) = a_n,$$

that is, if

$$\xi' = \eta',$$

$$(4) \qquad\qquad x_n = a_n/(1 - \lambda_n c) \qquad\qquad (\lambda_n \neq 1/c),$$
$$a_n = 0 \qquad\qquad (\lambda_n = 1/c).$$

The last equality gives a necessary and sufficient condition for a solution of (1), and (4) determines the solution. The values of x_n corresponding to those n for which $\lambda_n = 1/c$ remain arbitrary.

§61. Integral equations with symmetric kernel

The results presented in the preceding section can be applied to integral equations with symmetric kernel, that is, to equations of the form

$$(1) \qquad\qquad f(t) = \varphi(t) + \int_a^b K(t, s) f(s) \, ds,$$

where $K(t, s)$ satisfies the conditions

1) $K(t, s) = K(s, t)$,

2) $\displaystyle\int_a^b \int_a^b K^2(t, s) \, dt \, ds < \infty.$

The application of the results of §60 to equations of the form (1) is based on the following theorem:

THEOREM. *Let R be a space with measure μ. If a function $K(t, s)$ defined on $R^2 = R \times R$ satisfies the conditions*

$$(2) \qquad\qquad K(t, s) = K(s, t)$$

and

$$(3) \qquad\qquad \int_{R^2} K^2(t, s) \, d\mu^2 < \infty \qquad\qquad (\mu^2 = \mu \otimes \mu),$$

then the operator

$$g = Af$$

defined on $L_2(R, \mu)$ by the formula

$$g(t) = \int_R K(t, s) f(s) \, d\mu_s$$

is completely continuous and self-adjoint.

Proof. We shall denote the space $L_2(R, \mu)$ simply by L_2. Let $\{\psi_n(t)\}$ be a complete orthonormal set in L_2. The collection of all possible products $\psi_n(t)\psi_m(s)$ is a complete orthonormal set of functions in R^2 (see Ex. 3, §54), and

$$(4) \qquad\qquad K(t, s) = \sum_m \sum_n a_{mn} \psi_n(t) \psi_m(s)$$

in the mean [i.e., $\sum_m \sum_n$ converges to K in the norm of $L_2(R^2, \mu^2)$], where

$$a_{mn} = a_{nm}$$

(in view of (2)), and

$$\sum_m \sum_n a_{mn}^2 = \int_{R^2} K^2(t, s) \, d\mu^2 < \infty.$$

We set

(5) $$f(s) = \sum_n b_n \psi_n(s)$$

in the mean. Then

(6) $$g(x) = (Af)(x) = \sum_m \left(\sum_n a_{mn} b_n\right) \psi_m(x) = \sum_m c_m \psi_m(x)$$

in the mean. Also

$$c_m^2 = \left(\sum_{n=1}^\infty a_{mn} b_n\right)^2 \le \sum_{n=1}^\infty a_{mn}^2 \cdot \sum_{n=1}^\infty b_n^2 = \|f\|^2 \cdot a_m^2,$$

where

$$a_m^2 = \sum_n a_{mn}^2.$$

Since the series

$$\sum_{m=1}^\infty a_m^2 = \sum_m \sum_n a_{mn}^2$$

converges, for every $\epsilon > 0$ there is an m_0 such that

(7)
$$\sum_{m=m_0+1}^\infty a_m^2 < \epsilon,$$
$$\left\| g(x) - \sum_{m=1}^{m_0} c_m \psi_m(x) \right\|^2 = \sum_{m=m_0+1}^\infty c_m^2 < \epsilon \|f\|^2.$$

Now suppose that $\{f^{(k)}\}$ is weakly convergent to f. Then the corresponding $c_m^{(k)}$ converge to c_m for every m. Hence the sum

$$\sum_{m=1}^{m_0} c_m^{(k)} \psi_m(x)$$

converges in the mean to the sum

$$\sum_{m=1}^{m_0} c_m \psi_m(x)$$

for arbitrary fixed m_0. In view of the inequality (7) and the boundedness of the norm $\|f^{(k)}\|$ it follows that $\{g^{(k)}(x)\}$ (where $g^{(k)} = Af^{(k)}$) converges in the mean to $g(x)$. This proves that A is completely continuous. Multiplying (4) by (5), integrating with respect to μ_t and comparing the result with (6), we see that

$$(Af)(s) = \int_R K(s, t) f(t) \, d\mu_t.$$

This and Fubini's theorem imply that

$$(Af, g) = \int_R \left(\int_R K(s, t) f(t) \, d\mu_t \cdot g(s) \right) d\mu_s$$

$$= \int_R f(t) \left(\int_R K(s, t) g(s) \, d\mu_s \right) d\mu_t$$

$$= (f, Ag),$$

that is, A is self-adjoint. This proves the theorem.

Hence the solution of an integral equation with symmetric kernel satisfying conditions (2) and (3) reduces to finding the eigenfunctions and eigenvalues of the corresponding integral operator. The actual solution of the latter problem usually requires the use of some approximation method, but such methods are outside the scope of this book.

EXERCISES

1. Let $R = [a, b]$ be an interval on the real line, μ linear Lebesgue measure, $H = L_2(R, \mu)$, $K(t, s)$ as in the theorem of §61. Then by Theorem 1 of §59, the operator A determined by K has an orthonormal sequence of eigenfunctions $\{\varphi_n\}$ corresponding to a sequence $\{\lambda_n\}$ of eigenvalues $\lambda_n \neq 0$. Further, for $f(t) \in H$, $f(t) = \sum_{n=1}^{\infty} (f, \varphi_k)\varphi_k(t) + h(t)$ in the mean, with $Ah = 0$, i.e., $\int_R h(t) K(t, s) \, d\mu_s = 0$ and $g(s) = (Af)(s) = \sum_{k=1}^{\infty} \lambda_k (f, \varphi_k)\varphi_k(s)$ in the mean.

Suppose now that there is a constant M such that $\int_R |K(t, s)|^2 \, d\mu_s < M^2$ for all $t \in [a, b]$. An example is furnished by $K(t, s) = |t - s|^{-\alpha}$, $\alpha < \frac{1}{2}$.

Show that the series for $g(s)$ will converge uniformly and absolutely (pointwise) to $g(s)$.

Hints: For uniform convergence, apply the condition and Schwarz's inequality. For absolute convergence, show that mean convergence of a series with orthogonal terms is equivalent to convergence of the series with positive terms $(\sum_k \| \varphi_k \|^2)$ and that therefore the convergence is independent of the order of the terms.

SUPPLEMENT AND CORRECTIONS TO VOLUME 1

(1) p. 28, l. 23. Substitute $G_\alpha(x)$ for G_α.

(2) p. 46, l. 13* (l. 13 from bottom). Replace "the method of successive approximations is not applicable" by "the method of successive approximations is, in general, not applicable".

(3) p. 49, after l. 10* insert: "An arbitrary continuous function may be chosen for $f_0(x)$".

(4) p. 50, l. 1*. Replace the λ after the inequality sign by $|\lambda|$.

(5) p. 51, l. 2. Replace M by M^2.

p. 51, l. 3. Replace M by M^n (two times). Replace λ^n by $|\lambda|^n$ (two times).

p. 51, l. 5. Replace λ^n by $|\lambda|^n M^n$.

(6) p. 56, l. 6. Replace "closed region" by "closed bounded region".

p. 59, l. 2*. Replace the first occurrence of Y by X.

(7) p. 61, after l. 9 insert: "A mapping $y = f(x)$ is said to be *uniformly continuous* if for every $\epsilon > 0$ there is a $\delta > 0$ such that $\rho(f(x_1), f(x_2)) < \epsilon$ for all x_1, x_2 for which $\rho(x_1, x_2) < \delta$. The following theorem holds: *Every continuous mapping of a compactum into a compactum is uniformly continuous*. This theorem is proved in the same way as the uniform continuity of a function continuous on a closed interval".

(8) p. 61, l. 18. After "Proof" insert: "We shall prove the necessity first. If D is compact, D contains a finite $\epsilon/3$-chain $f_1, \cdots, f_N$. Since each mapping f_i is continuous, it is uniformly continuous. Therefore, there is a $\delta > 0$ such that

$$\rho(f_i(x_1), f_i(x_2)) < \epsilon/3 \qquad (1 \le i \le n)$$

if

$$\rho(x_1, x_2) < \delta.$$

If $f \in D$, there exists an f_i such that

$$\rho(f, f_i) < \epsilon/3.$$

Then

$$\rho(f(x_1), f(x_2)) \le \rho(f(x_1), f_i(x_1)) + \rho(f_i(x_1), f_i(x_2))$$
$$+ \rho(f_i(x_2), f(x_2)) < \epsilon/3 + \epsilon/3 + \epsilon/3 = \epsilon$$

if $\rho(x_1, x_2) < \delta$. But this means that the set of all $f \in D$ is equicontinuous. We shall now prove the sufficiency."

(9) p. 72, l. 3*. Replace "max" by "sup".

(10) p. 77, l. 9. Replace "continuous" by "continuous at a point x_0".

p. 77, l. 13. Replace $\| x_1 - x_2 \|$ by $\| x - x_0 \|$.

p. 77, l. 11. Replace $| f(x_1) - f(x_2) |$ by $| f(x) - f(x_0) |$.

p. 77, l. 20. Replace "continuous" by "uniformly continuous". Delete "everywhere in R".

(11) p. 80, l. 9*. Replace "$x_0 \notin L_f$" by "x_0 is a fixed element of the complement of L_f".

(12) p. 84, l. 12. Replace "$\sup_n x_n$" by "$\sup_n | x_n |$".

(13) p. 92, l. 14*, 13*. The assertion that the functionals δ_{t_0} generate a dense subset of $\bar{C}$ is not true. Replace "satisfies the conditions of Theorem 1, i.e. linear combinations of these functionals are everywhere dense in $\bar{C}_{[a,b]}$" by "has the property that if a sequence $\{x_n(t)\}$ is bounded and $\varphi(x_n) \to \varphi(x)$ for all $\varphi \in \Delta$, then $\{x_n(t)\}$ is weakly convergent to $x(t)$".

(14) p. 94, l. 9* ff. The metric introduced here leads to a convergence which is equivalent to the weak convergence of functionals in every bounded subset of $\bar{R}$ (but not in all of $\bar{R}$). In l. 6*, after "so that" insert "in every bounded subset of $\bar{R}$". On p. 95, l. 10, after "that" insert "for bounded sequences of $\bar{R}$".

(15) p. 116. The proof of Theorem 5 contains an error. It should be replaced by the following:

Proof. 1°. We note first that every nonvanishing eigenvalue of a completely continuous operator has finite multiplicity. In fact, the set E_λ of all eigenvectors corresponding to an eigenvalue λ is a linear subspace whose dimension is equal to the multiplicity of the eigenvalue. If this subspace were infinite-dimensional for some $\lambda \neq 0$, the operator A would not be completely continuous in E_λ, and hence would not be completely continuous in the whole space.

2°. Now to complete the proof of the theorem it remains to show that if $\{\lambda_n\}$ is a sequence of distinct eigenvalues of a completely continuous operator A, then $\lambda_n \to 0$ as $n \to \infty$. Let x_n be an eigenvector of A corresponding to the eigenvalue λ_n. The vectors x_n are linearly independent. Let E_n ($n = 1, 2, \cdots$) be the subspace of all the elements of the form

$$y = \sum_{i=1}^{n} \alpha_i x_i .$$

For each $y \in E_n$,

$$y - \lambda_n^{-1} A y = \sum_{i=1}^{n} \alpha_i x_i - \sum_{i=1}^{n} \alpha_i \lambda_i \lambda_n^{-1} x_i = \sum_{i=1}^{n-1} (1 - \lambda_i \lambda_n^{-1}) \alpha_i x_i ,$$

whence it is clear that $y - \lambda_n^{-1} A y \in E_{n-1}$.

Choose a sequence $\{y_n\}$ such that

$$y_n \in E_n , \qquad \| y_n \| = 1 , \qquad \rho(y_n , E_{n-1}) > \tfrac{1}{2} .$$

(The existence of such a sequence was proved on p. 118, l. 6 ff.)

We now suppose that the sequence $1/\lambda_n$ is bounded. Then the set $\{A(y_n/\lambda_n)\}$ is compact. But this is impossible, since

$$\| A(y_p/\lambda_p) - A(y_q/\lambda_q) \|$$

$$= \| y_p - [y_p - \lambda_p^{-1}Ay_p + A(y_q/\lambda_q)] \| > \tfrac{1}{2} \quad (p > q),$$

inasmuch as $y_p - \lambda_p^{-1}Ay_p + A(y_q/\lambda_q) \in E_{p-1}$. This contradiction proves the assertion.

(16) p. 119, l. 12. The assertion that G_0 is a subspace is true, but not obvious. Therefore the sentence "Let G_0 be the subspace consisting of all elements of the form $x - Ax$" should be replaced by the following: "Let G_0 be the linear manifold consisting of all elements of the form $x - Ax$. We shall show that G_0 is closed. Let $\tilde{T}$ be a one-to-one mapping of the quotient space E/N (where N is the subspace of the elements satisfying the condition $x - Ax = 0$) onto G_0. (For the definition of quotient space see Ex. 5, §57.) We must show that the inverse mapping $\tilde{T}^{-1}$ is continuous. It is sufficient to show that it is continuous at $y = 0$. Suppose that this is not so; then there exists a sequence $y_n \to 0$ such that $\| \xi_n \| \geq \rho > 0$, where $\xi_n = \tilde{T}^{-1}y_n$. Setting $\eta_n = \xi_n/\| \xi_n \|$ and $z_n = y_n/\| \xi_n \|$, we obtain a sequence $\{\eta_n\}$ satisfying the conditions:

$$\| \eta_n \| = 1, \qquad \tilde{T}\eta_n = z_n \to 0.$$

If we choose in each class η_n a representative x_n such that $\| x_n \| \leq 2$, we obtain a bounded sequence, and $z_n = Tx_n = x_n - Ax_n \to 0$. But since the operator A is completely continuous, $\{Ax_n\}$ contains a fundamental subsequence $\{Ax_{n_k}\}$. The sequence $x_p = z_p + Ax_p$ (where $x_p = x_{n_p}$ and $z_p = z_{n_p}$) is also fundamental and therefore converges to an element x_0. Hence $z_p = Tx_p \to Tx_0$, so that $Tx_0 = 0$, that is, $x_0 \in N$. But then $\| \eta_p \| \leq \| x_p - x_0 \| \to 0$, which contradicts the condition $\| \eta_p \| = 1$. This contradiction proves the continuity of $\tilde{T}^{-1}$ and shows that G_0 is closed. Hence G_0 is a subspace".

INDEX

A CATALOG OF SELECTED
DOVER BOOKS
IN SCIENCE AND MATHEMATICS

Astronomy

BURNHAM'S CELESTIAL HANDBOOK, Robert Burnham, Jr. Thorough guide to the stars beyond our solar system. Exhaustive treatment. Alphabetical by constellation: Andromeda to Cetus in Vol. 1; Chamaeleon to Orion in Vol. 2; and Pavo to Vulpecula in Vol. 3. Hundreds of illustrations. Index in Vol. 3. 2,000pp. 6⅛ x 9¼.

Vol. I: 0-486-23567-X
Vol. II: 0-486-23568-8
Vol. III: 0-486-23673-0

EXPLORING THE MOON THROUGH BINOCULARS AND SMALL TELESCOPES, Ernest H. Cherrington, Jr. Informative, profusely illustrated guide to locating and identifying craters, rills, seas, mountains, other lunar features. Newly revised and updated with special section of new photos. Over 100 photos and diagrams. 240pp. 8¼ x 11. 0-486-24491-1

THE EXTRATERRESTRIAL LIFE DEBATE, 1750–1900, Michael J. Crowe. First detailed, scholarly study in English of the many ideas that developed from 1750 to 1900 regarding the existence of intelligent extraterrestrial life. Examines ideas of Kant, Herschel, Voltaire, Percival Lowell, many other scientists and thinkers. 16 illustrations. 704pp. 5⅜ x 8½. 0-486-40675-X

THEORIES OF THE WORLD FROM ANTIQUITY TO THE COPERNICAN REVOLUTION, Michael J. Crowe. Newly revised edition of an accessible, enlightening book re-creates the change from an earth-centered to a sun-centered conception of the solar system. 242pp. 5⅜ x 8½. 0-486-41444-2

ARISTARCHUS OF SAMOS: The Ancient Copernicus, Sir Thomas Heath. Heath's history of astronomy ranges from Homer and Hesiod to Aristarchus and includes quotes from numerous thinkers, compilers, and scholasticists from Thales and Anaximander through Pythagoras, Plato, Aristotle, and Heraclides. 34 figures. 448pp. 5⅜ x 8½. 0-486-43886-4

A COMPLETE MANUAL OF AMATEUR ASTRONOMY: TOOLS AND TECHNIQUES FOR ASTRONOMICAL OBSERVATIONS, P. Clay Sherrod with Thomas L. Koed. Concise, highly readable book discusses: selecting, setting up and maintaining a telescope; amateur studies of the sun; lunar topography and occultations; observations of Mars, Jupiter, Saturn, the minor planets and the stars; an introduction to photoelectric photometry; more. 1981 ed. 124 figures. 25 halftones. 37 tables. 335pp. 6½ x 9¼. 0-486-42820-8

AMATEUR ASTRONOMER'S HANDBOOK, J. B. Sidgwick. Timeless, comprehensive coverage of telescopes, mirrors, lenses, mountings, telescope drives, micrometers, spectroscopes, more. 189 illustrations. 576pp. 5⅝ x 8¼. (Available in U.S. only.) 0-486-24034-7

STAR LORE: Myths, Legends, and Facts, William Tyler Olcott. Captivating retellings of the origins and histories of ancient star groups include Pegasus, Ursa Major, Pleiades, signs of the zodiac, and other constellations. "Classic."—Sky & Telescope. 58 illustrations. 544pp. 5⅜ x 8½. 0-486-43581-4

Chemistry

THE SCEPTICAL CHYMIST: THE CLASSIC 1661 TEXT, Robert Boyle. Boyle defines the term "element," asserting that all natural phenomena can be explained by the motion and organization of primary particles. 1911 ed. viii+232pp. 5⅜ x 8½. 0-486-42825-7

RADIOACTIVE SUBSTANCES, Marie Curie. Here is the celebrated scientist's doctoral thesis, the prelude to her receipt of the 1903 Nobel Prize. Curie discusses establishing atomic character of radioactivity found in compounds of uranium and thorium; extraction from pitchblende of polonium and radium; isolation of pure radium chloride; determination of atomic weight of radium; plus electric, photographic, luminous, heat, color effects of radioactivity. ii+94pp. 5⅜ x 8½. 0-486-42550-9

CHEMICAL MAGIC, Leonard A. Ford. Second Edition, Revised by E. Winston Grundmeier. Over 100 unusual stunts demonstrating cold fire, dust explosions, much more. Text explains scientific principles and stresses safety precautions. 128pp. 5⅜ x 8½. 0-486-67628-5

MOLECULAR THEORY OF CAPILLARITY, J. S. Rowlinson and B. Widom. History of surface phenomena offers critical and detailed examination and assessment of modern theories, focusing on statistical mechanics and application of results in mean-field approximation to model systems. 1989 edition. 352pp. 5⅜ x 8½. 0-486-42544-4

CHEMICAL AND CATALYTIC REACTION ENGINEERING, James J. Carberry. Designed to offer background for managing chemical reactions, this text examines behavior of chemical reactions and reactors; fluid-fluid and fluid-solid reaction systems; heterogeneous catalysis and catalytic kinetics; more. 1976 edition. 672pp. 6⅛ x 9¼. 0-486-41736-0 $31.95

ELEMENTS OF CHEMISTRY, Antoine Lavoisier. Monumental classic by founder of modern chemistry in remarkable reprint of rare 1790 Kerr translation. A must for every student of chemistry or the history of science. 539pp. 5⅜ x 8½. 0-486-64624-6

MOLECULES AND RADIATION: An Introduction to Modern Molecular Spectroscopy. Second Edition, Jeffrey I. Steinfeld. This unified treatment introduces upper-level undergraduates and graduate students to the concepts and the methods of molecular spectroscopy and applications to quantum electronics, lasers, and related optical phenomena. 1985 edition. 512pp. 5⅜ x 8½. 0-486-44152-0

A SHORT HISTORY OF CHEMISTRY, J. R. Partington. Classic exposition explores origins of chemistry, alchemy, early medical chemistry, nature of atmosphere, theory of valency, laws and structure of atomic theory, much more. 428pp. 5⅜ x 8½. (Available in U.S. only.) 0-486-65977-1

GENERAL CHEMISTRY, Linus Pauling. Revised 3rd edition of classic first-year text by Nobel laureate. Atomic and molecular structure, quantum mechanics, statistical mechanics, thermodynamics correlated with descriptive chemistry. Problems. 992pp. 5⅜ x 8½. 0-486-65622-5

ELECTRON CORRELATION IN MOLECULES, S. Wilson. This text addresses one of theoretical chemistry's central problems. Topics include molecular electronic structure, independent electron models, electron correlation, the linked diagram theorem, and related topics. 1984 edition. 304pp. 5⅜ x 8½. 0-486-45879-2

Engineering

DE RE METALLICA, Georgius Agricola. The famous Hoover translation of greatest treatise on technological chemistry, engineering, geology, mining of early modern times (1556). All 289 original woodcuts. 638pp. 6¾ x 11.   0-486-60006-8

FUNDAMENTALS OF ASTRODYNAMICS, Roger Bate et al. Modern approach developed by U.S. Air Force Academy. Designed as a first course. Problems, exercises. Numerous illustrations. 455pp. 5⅜ x 8½.   0-486-60061-0

DYNAMICS OF FLUIDS IN POROUS MEDIA, Jacob Bear. For advanced students of ground water hydrology, soil mechanics and physics, drainage and irrigation engineering and more. 335 illustrations. Exercises, with answers. 784pp. 6⅛ x 9¼.

0-486-65675-6

THEORY OF VISCOELASTICITY (SECOND EDITION), Richard M. Christensen. Complete consistent description of the linear theory of the viscoelastic behavior of materials. Problem-solving techniques discussed. 1982 edition. 29 figures. xiv+364pp. 6⅛ x 9¼.   0-486-42880-X

MECHANICS, J. P. Den Hartog. A classic introductory text or refresher. Hundreds of applications and design problems illuminate fundamentals of trusses, loaded beams and cables, etc. 334 answered problems. 462pp. 5⅜ x 8½.   0-486-60754-2

MECHANICAL VIBRATIONS, J. P. Den Hartog. Classic textbook offers lucid explanations and illustrative models, applying theories of vibrations to a variety of practical industrial engineering problems. Numerous figures. 233 problems, solutions. Appendix. Index. Preface. 436pp. 5⅜ x 8½.   0-486-64785-4

STRENGTH OF MATERIALS, J. P. Den Hartog. Full, clear treatment of basic material (tension, torsion, bending, etc.) plus advanced material on engineering methods, applications. 350 answered problems. 323pp. 5⅜ x 8½.   0-486-60755-0

A HISTORY OF MECHANICS, René Dugas. Monumental study of mechanical principles from antiquity to quantum mechanics. Contributions of ancient Greeks, Galileo, Leonardo, Kepler, Lagrange, many others. 671pp. 5⅜ x 8½. 0-486-65632-2

STABILITY THEORY AND ITS APPLICATIONS TO STRUCTURAL MECHANICS, Clive L. Dym. Self-contained text focuses on Koiter postbuckling analyses, with mathematical notions of stability of motion. Basing minimum energy principles for static stability upon dynamic concepts of stability of motion, it develops asymptotic buckling and postbuckling analyses from potential energy considerations, with applications to columns, plates, and arches. 1974 ed. 208pp. 5⅜ x 8½.
0-486-42541-X

BASIC ELECTRICITY, U.S. Bureau of Naval Personnel. Originally a training course; best nontechnical coverage. Topics include batteries, circuits, conductors, AC and DC, inductance and capacitance, generators, motors, transformers, amplifiers, etc. Many questions with answers. 349 illustrations. 1969 edition. 448pp. 6½ x

ROCKETS, Robert Goddard. Two of the most significant publications in the history of rocketry and jet propulsion: "A Method of Reaching Extreme Altitudes" (1919) and "Liquid Propellant Rocket Development" (1936). 128pp. 5⅜ x 8½. 0-486-42537-1

STATISTICAL MECHANICS: PRINCIPLES AND APPLICATIONS, Terrell L. Hill. Standard text covers fundamentals of statistical mechanics, applications to fluctuation theory, imperfect gases, distribution functions, more. 448pp. 5⅜ x 8½.
0-486-65390-0

ENGINEERING AND TECHNOLOGY 1650–1750: ILLUSTRATIONS AND TEXTS FROM ORIGINAL SOURCES, Martin Jensen. Highly readable text with more than 200 contemporary drawings and detailed engravings of engineering projects dealing with surveying, leveling, materials, hand tools, lifting equipment, transport and erection, piling, bailing, water supply, hydraulic engineering, and more. Among the specific projects outlined-transporting a 50-ton stone to the Louvre, erecting an obelisk, building timber locks, and dredging canals. 207pp. 8⅜ x 11¼.
0-486-42232-1

THE VARIATIONAL PRINCIPLES OF MECHANICS, Cornelius Lanczos. Graduate level coverage of calculus of variations, equations of motion, relativistic mechanics, more. First inexpensive paperbound edition of classic treatise. Index. Bibliography. 418pp. 5⅜ x 8½. 0-486-65067-7

PROTECTION OF ELECTRONIC CIRCUITS FROM OVERVOLTAGES, Ronald B. Standler. Five-part treatment presents practical rules and strategies for circuits designed to protect electronic systems from damage by transient overvoltages. 1989 ed. xxiv+434pp. 6⅛ x 9¼. 0-486-42552-5

ROTARY WING AERODYNAMICS, W. Z. Stepniewski. Clear, concise text covers aerodynamic phenomena of the rotor and offers guidelines for helicopter performance evaluation. Originally prepared for NASA. 537 figures. 640pp. 6⅛ x 9¼.
0-486-64647-5

INTRODUCTION TO SPACE DYNAMICS, William Tyrrell Thomson. Comprehensive, classic introduction to space-flight engineering for advanced undergraduate and graduate students. Includes vector algebra, kinematics, transformation of coordinates. Bibliography. Index. 352pp. 5⅜ x 8½. 0-486-65113-4

HISTORY OF STRENGTH OF MATERIALS, Stephen P. Timoshenko. Excellent historical survey of the strength of materials with many references to the theories of elasticity and structure. 245 figures. 452pp. 5⅜ x 8½. 0-486-61187-6

ANALYTICAL FRACTURE MECHANICS, David J. Unger. Self-contained text supplements standard fracture mechanics texts by focusing on analytical methods for determining crack-tip stress and strain fields. 336pp. 6⅛ x 9¼. 0-486-41737-9

STATISTICAL MECHANICS OF ELASTICITY, J. H. Weiner. Advanced, self-contained treatment illustrates general principles and elastic behavior of solids. Part 1, based on classical mechanics, studies thermoelastic behavior of crystalline and polymeric solids. Part 2, based on quantum mechanics, focuses on interatomic force laws, behavior of solids, and thermally activated processes. For students of physics and chemistry and for polymer physicists. 1983 ed. 96 figures. 496pp. 5⅜ x 8½. 0-486-42260-7

Mathematics

FUNCTIONAL ANALYSIS (Second Corrected Edition), George Bachman and Lawrence Narici. Excellent treatment of subject geared toward students with background in linear algebra, advanced calculus, physics and engineering. Text covers introduction to inner-product spaces, normed, metric spaces, and topological spaces; complete orthonormal sets, the Hahn-Banach Theorem and its consequences, and many other related subjects. 1966 ed. 544pp. 6⅛ x 9¼. 0-486-40251-7

DIFFERENTIAL MANIFOLDS, Antoni A. Kosinski. Introductory text for advanced undergraduates and graduate students presents systematic study of the topological structure of smooth manifolds, starting with elements of theory and concluding with method of surgery. 1993 edition. 288pp. 5⅜ x 8½. 0-486-46244-7

VECTOR AND TENSOR ANALYSIS WITH APPLICATIONS, A. I. Borisenko and I. E. Tarapov. Concise introduction. Worked-out problems, solutions, exercises. 257pp. 5⅝ x 8¼. 0-486-63833-2

AN INTRODUCTION TO ORDINARY DIFFERENTIAL EQUATIONS, Earl A. Coddington. A thorough and systematic first course in elementary differential equations for undergraduates in mathematics and science, with many exercises and problems (with answers). Index. 304pp. 5⅜ x 8½. 0-486-65942-9

FOURIER SERIES AND ORTHOGONAL FUNCTIONS, Harry F. Davis. An incisive text combining theory and practical example to introduce Fourier series, orthogonal functions and applications of the Fourier method to boundary-value problems. 570 exercises. Answers and notes. 416pp. 5⅜ x 8½. 0-486-65973-9

COMPUTABILITY AND UNSOLVABILITY, Martin Davis. Classic graduate-level introduction to theory of computability, usually referred to as theory of recurrent functions. New preface and appendix. 288pp. 5⅜ x 8½. 0-486-61471-9

AN INTRODUCTION TO MATHEMATICAL ANALYSIS, Robert A. Rankin. Dealing chiefly with functions of a single real variable, this text by a distinguished educator introduces limits, continuity, differentiability, integration, convergence of infinite series, double series, and infinite products. 1963 edition. 624pp. 5⅜ x 8½. 0-486-46251-X

METHODS OF NUMERICAL INTEGRATION (SECOND EDITION), Philip J. Davis and Philip Rabinowitz. Requiring only a background in calculus, this text covers approximate integration over finite and infinite intervals, error analysis, approximate integration in two or more dimensions, and automatic integration. 1984 edition. 624pp. 5⅜ x 8½. 0-486-45339-1

INTRODUCTION TO LINEAR ALGEBRA AND DIFFERENTIAL EQUATIONS, John W. Dettman. Excellent text covers complex numbers, determinants, orthonormal bases, Laplace transforms, much more. Exercises with solutions. Undergraduate level. 416pp. 5⅜ x 8½. 0-486-65191-6

RIEMANN'S ZETA FUNCTION, H. M. Edwards. Superb, high-level study of landmark 1859 publication entitled "On the Number of Primes Less Than a Given Magnitude" traces developments in mathematical theory that it inspired. xiv+315pp. 5⅜ x 8½. 0-486-41740-9

CALCULUS OF VARIATIONS WITH APPLICATIONS, George M. Ewing. Applications-oriented introduction to variational theory develops insight and promotes understanding of specialized books, research papers. Suitable for advanced undergraduate/graduate students as primary, supplementary text. 352pp. 5⅜ x 8½. 0-486-64856-7

MATHEMATICIAN'S DELIGHT, W. W. Sawyer. "Recommended with confidence" by *The Times Literary Supplement*, this lively survey was written by a renowned teacher. It starts with arithmetic and algebra, gradually proceeding to trigonometry and calculus. 1943 edition. 240pp. 5⅜ x 8½. 0-486-46240-4

ADVANCED EUCLIDEAN GEOMETRY, Roger A. Johnson. This classic text explores the geometry of the triangle and the circle, concentrating on extensions of Euclidean theory, and examining in detail many relatively recent theorems. 1929 edition. 336pp. 5⅜ x 8½. 0-486-46237-4

COUNTEREXAMPLES IN ANALYSIS, Bernard R. Gelbaum and John M. H. Olmsted. These counterexamples deal mostly with the part of analysis known as "real variables." The first half covers the real number system, and the second half encompasses higher dimensions. 1962 edition. xxiv+198pp. 5⅜ x 8½.0-486-42875-3

CATASTROPHE THEORY FOR SCIENTISTS AND ENGINEERS, Robert Gilmore. Advanced-level treatment describes mathematics of theory grounded in the work of Poincaré, R. Thom, other mathematicians. Also important applications to problems in mathematics, physics, chemistry and engineering. 1981 edition. References. 28 tables. 397 black-and-white illustrations. xvii + 666pp. 6⅛ x 9¼. 0-486-67539-4

COMPLEX VARIABLES: Second Edition, Robert B. Ash and W. P. Novinger. Suitable for advanced undergraduates and graduate students, this newly revised treatment covers Cauchy theorem and its applications, analytic functions, and the prime number theorem. Numerous problems and solutions. 2004 edition. 224pp. 6½ x 9¼. 0-486-46250-1

NUMERICAL METHODS FOR SCIENTISTS AND ENGINEERS, Richard Hamming. Classic text stresses frequency approach in coverage of algorithms, polynomial approximation, Fourier approximation, exponential approximation, other topics. Revised and enlarged 2nd edition. 721pp. 5⅜ x 8½. 0-486-65241-6

INTRODUCTION TO NUMERICAL ANALYSIS (2nd Edition), F. B. Hildebrand. Classic, fundamental treatment covers computation, approximation, interpolation, numerical differentiation and integration, other topics. 150 new problems. 669pp. 5⅜ x 8½. 0-486-65363-3

MARKOV PROCESSES AND POTENTIAL THEORY, Robert M. Blumental and Ronald K. Getoor. This graduate-level text explores the relationship between Markov processes and potential theory in terms of excessive functions, multiplicative functionals and subprocesses, additive functionals and their potentials, and dual processes. 1968 edition. 320pp. 5⅜ x 8½.   0-486-46263-3

ABSTRACT SETS AND FINITE ORDINALS: An Introduction to the Study of Set Theory, G. B. Keene. This text unites logical and philosophical aspects of set theory in a manner intelligible to mathematicians without training in formal logic and to logicians without a mathematical background. 1961 edition. 112pp. 5⅜ x 8½.
0-486-46249-8

INTRODUCTORY REAL ANALYSIS, A.N. Kolmogorov, S. V. Fomin. Translated by Richard A. Silverman. Self-contained, evenly paced introduction to real and functional analysis. Some 350 problems. 403pp. 5⅜ x 8½.   0-486-61226-0

APPLIED ANALYSIS, Cornelius Lanczos. Classic work on analysis and design of finite processes for approximating solution of analytical problems. Algebraic equations, matrices, harmonic analysis, quadrature methods, much more. 559pp. 5⅜ x 8½.   0-486-65656-X

AN INTRODUCTION TO ALGEBRAIC STRUCTURES, Joseph Landin. Superb self-contained text covers "abstract algebra": sets and numbers, theory of groups, theory of rings, much more. Numerous well-chosen examples, exercises. 247pp. 5⅜ x 8½.   0-486-65940-2

QUALITATIVE THEORY OF DIFFERENTIAL EQUATIONS, V. V. Nemytskii and V.V. Stepanov. Classic graduate-level text by two prominent Soviet mathematicians covers classical differential equations as well as topological dynamics and ergodic theory. Bibliographies. 523pp. 5⅜ x 8½.   0-486-65954-2

THEORY OF MATRICES, Sam Perlis. Outstanding text covering rank, nonsingularity and inverses in connection with the development of canonical matrices under the relation of equivalence, and without the intervention of determinants. Includes exercises. 237pp. 5⅜ x 8½.   0-486-66810-X

INTRODUCTION TO ANALYSIS, Maxwell Rosenlicht. Unusually clear, accessible coverage of set theory, real number system, metric spaces, continuous functions, Riemann integration, multiple integrals, more. Wide range of problems. Undergraduate level. Bibliography. 254pp. 5⅜ x 8½.   0-486-65038-3

MODERN NONLINEAR EQUATIONS, Thomas L. Saaty. Emphasizes practical solution of problems; covers seven types of equations. ". . . a welcome contribution to the existing literature. . . ."—*Math Reviews.* 490pp. 5⅜ x 8½.   0-486-64232-1

MATRICES AND LINEAR ALGEBRA, Hans Schneider and George Phillip Barker. Basic textbook covers theory of matrices and its applications to systems of linear equations and related topics such as determinants, eigenvalues and differential equations. Numerous exercises. 432pp. 5⅜ x 8½.   0-486-66014-1

LINEAR ALGEBRA, Georgi E. Shilov. Determinants, linear spaces, matrix algebras, similar topics. For advanced undergraduates, graduates. Silverman translation. 387pp. 5⅜ x 8½.   0-486-63518-X

MATHEMATICAL METHODS OF GAME AND ECONOMIC THEORY:

Revised Edition, Jean-Pierre Aubin. This text begins with optimization theory and convex analysis, followed by topics in game theory and mathematical economics, and concluding with an introduction to nonlinear analysis and control theory. 1982 edition. 656pp. 6⅛ x 9¼. 0-486-46265-X

SET THEORY AND LOGIC, Robert R. Stoll. Lucid introduction to unified theory of mathematical concepts. Set theory and logic seen as tools for conceptual understanding of real number system. 496pp. 5⅜ x 8¼. 0-486-63829-4

TENSOR CALCULUS, J.L. Synge and A. Schild. Widely used introductory text covers spaces and tensors, basic operations in Riemannian space, non-Riemannian spaces, etc. 324pp. 5⅜ x 8¼. 0-486-63612-7

ORDINARY DIFFERENTIAL EQUATIONS, Morris Tenenbaum and Harry Pollard. Exhaustive survey of ordinary differential equations for undergraduates in mathematics, engineering, science. Thorough analysis of theorems. Diagrams. Bibliography. Index. 818pp. 5⅜ x 8½. 0-486-64940-7

INTEGRAL EQUATIONS, F. G. Tricomi. Authoritative, well-written treatment of extremely useful mathematical tool with wide applications. Volterra Equations, Fredholm Equations, much more. Advanced undergraduate to graduate level. Exercises. Bibliography. 238pp. 5⅜ x 8½. 0-486-64828-1

FOURIER SERIES, Georgi P. Tolstov. Translated by Richard A. Silverman. A valuable addition to the literature on the subject, moving clearly from subject to subject and theorem to theorem. 107 problems, answers. 336pp. 5⅜ x 8½. 0-486-63317-9

INTRODUCTION TO MATHEMATICAL THINKING, Friedrich Waismann. Examinations of arithmetic, geometry, and theory of integers; rational and natural numbers; complete induction; limit and point of accumulation; remarkable curves; complex and hypercomplex numbers, more. 1959 ed. 27 figures. xii+260pp. 5⅜ x 8½. 0-486-42804-8

THE RADON TRANSFORM AND SOME OF ITS APPLICATIONS, Stanley R. Deans. Of value to mathematicians, physicists, and engineers, this excellent introduction covers both theory and applications, including a rich array of examples and literature. Revised and updated by the author. 1993 edition. 304pp. 6⅛ x 9¼.
0-486-46241-2

CALCULUS OF VARIATIONS, Robert Weinstock. Basic introduction covering isoperimetric problems, theory of elasticity, quantum mechanics, electrostatics, etc. Exercises throughout. 326pp. 5⅜ x 8½. 0-486-63069-2

THE CONTINUUM: A CRITICAL EXAMINATION OF THE FOUNDATION OF ANALYSIS, Hermann Weyl. Classic of 20th-century foundational research deals with the conceptual problem posed by the continuum. 156pp. 5⅜ x 8½.
0-486-67982-9

CHALLENGING MATHEMATICAL PROBLEMS WITH ELEMENTARY SOLUTIONS, A. M. Yaglom and I. M. Yaglom. Over 170 challenging problems on probability theory, combinatorial analysis, points and lines, topology, convex polygons, many other topics. Solutions. Total of 445pp. 5⅜ x 8½. Two-vol. set.
Vol. I: 0-486-65536-9   Vol. II: 0-486-65537-7

INTRODUCTION TO PARTIAL DIFFERENTIAL EQUATIONS WITH